LONGMAN
hOMEWORK **h**ANDBOOKS

GCSE

PHYSICS

Keith Palfreyman

LONGMAN

HOMEWORK HANDBOOKS

Series editors:

Geoff Black and Stuart Wall

Other titles in this series:
BIOLOGY
CHEMISTRY
GEOGRAPHY
MATHEMATICS
SCIENCE

Addison Wesley Longman Limited
Edinburgh Gate, Harlow
Essex CM20 2JE, England
and Associated Companies throughout the world

© Addison Wesley Longman Limited 1999

The right of John Sadler & Mark McElroy to be identified as authors of this work has been asserted by them in accordance with the Copyright, Designs and Patents Act 1988

First published 1999

British Library Cataloguing in Publication Data
A catalogue entry for this title is available from the British Library

ISBN 0-582-36917-7

Set by 3 in Stone 9½/11½
Produced by Addison Wesley Longman Singapore (Pte) Ltd
Printed in Singapore

CONTENTS

ACKNOWLEDGEMENTS

Once again my thanks to all those who helped me to check over this material and to those who commented on previous work. My special thanks to Angela for her support and to my sons Paul and Kevin for their reading comments. The helpful attitude of the series editors, the copy readers and others involved in the process is also much appreciated.

Keith Palfreyman

WHAT IS A HOMEWORK HANDBOOK?

This Homework Handbook is a resource. It is for you, the student, to decide exactly how you use the book. The main purpose of this book is to help you when you are working by yourself – when you're researching or revising. It is designed to be extremely flexible, both in content and organization.

What's in this book

The Homework Handbook consists of hundreds of important physical terms, listed alphabetically.

- **Definitions, explanations and examples:** These terms all relate to key ideas in the most recent GCSE syllabuses. Each idea or topic is defined and concisely explained, and often incorporates appropriate examples.

- **Remember bubbles:** Many topics include extra hints, tips and comments, which may reinforce learning, highlight how different topics relate or let you know of common pitfalls.

- **Checkpoint questions:** Many of the topics include 'Checkpoints' – these are questions based on recent GCSE examination questions.

- **Checkpoint answers:** Once you have had a go at answering a Checkpoint, you can turn to the answers at the back of the book to see how you got on. The answers include Examiner's tips on what to look out for when answering similar questions, and cross-references to related topics.

- **Cross-references:** Most topics are cross-referenced to other topics. This allows you to broaden and deepen your understanding of related themes. The cross-references also give you a chance to go back over the basic principles of themes with which you are having difficulty. Cross-references are easily identifiable:

- Either as words in bold italic type within the text:

  An ***electromotive force*** is produced . . .

- Or as words following the 'compass' symbol at the end of an entry:

  ✛ ***Dynamo, Transformer***

Using this book

- **Researching:** You can look up a particular term or topic just by finding it in the alphabetical list. Once you have the information you need, you can stop there; or you can follow up the related topics by using the cross-references; or you can try a Checkpoint question, to see if you have fully understood the topic.

- **Revising:** For revision, you can really test your understanding of GCSE Physics by dipping into the Homework Handbook to check out terms and topics you are revising or have already revised.

GOOD LUCK!

MAKING THE MOST OF HOMEWORK AND REVISION

Why do you need to do homework anyway?

There are lots of reasons for doing homework. Each of these will apply to you at some stage, depending on the topic you are learning and how close you are to the examination:

- **Homework is your own work:** Homework allows you to work at your own speed, and often in your own place and at a time that is convenient to you.

- **Homework is a chance to catch up with work:** You need to catch up if you have missed some work, or if you are having difficulty understanding a topic.

- **Homework is a chance to follow up:** If you want to know more about a particular topic, or you are doing a project, you can use homework to deepen your understanding, or just to generate a few more ideas.

- **Homework is a chance to prepare:** You can get more out of your class time if you are prepared. You can also be ready to ask, or answer, questions more effectively.

- **Homework is revision time:** You will need to do more work on your own as the examinations approach. In particular, you need to gain experience in answering the sort of questions you will meet in the examination.

- **Homework develops useful skills:** Doing homework encourages useful (and sometimes difficult to maintain!) habits, like concentration, organization and self-discipline.

Planning and doing your homework and revision

The best (and probably only) way of doing homework is to work to a realistic but challenging plan. What do you need in order to achieve this? The most successful homework strategy is the one that works best for you. Here are some basic suggestions:

- **You need motivation:** It's easier and more effective (and more enjoyable) if your motivation is high. In other words, you should have a clear purpose, and be keen to achieve it. There are certain rewards that help you develop motivation. Examples are getting higher marks and understanding the subject better.

- **You need a weekly homework timetable:** Write out a weekly plan of your homework tasks. This plan may already be decided by deadlines for each subject. If you need to make your own plan,

do a trial week first, to check that your workload is not too heavy (or too light).

- **You need to vary your homework:** Homework can consist of a series of different topics or subjects. This should keep you lively, alert and motivated.

- **You need to plan individual homework sessions for each topic or subject:** Each session should consist of fairly intensive work on a particular topic or subject. The session should not be too long (e.g. 30 minutes) and should have definite goals. Allow a break (e.g. five minutes) between sessions.

- **You need definite goals for each session:** For instance, you might have a specific piece of work which was set in class. You might want to re-read and add to your notes on a specific topic or practise answering a related examination question.

- **You need to be realistic:** Setting small, definite tasks increases your chances of completing them successfully. This should help with your confidence and motivation.

- **You need to be challenged:** If your task is too easy, you won't achieve much. Homework is a good example of 'no pain, no gain' – but that doesn't mean that homework has to be a chore!

- **You need to develop regular work habits:** This will avoid time-wasting periods of indecision, uncertainty and worry.

- **You need to adopt an active, not passive approach:** Learning is most effective when you are doing something. Summarizing information or answering a question is active. Reading through familiar notes can be quite passive, and possibly not very helpful (though it may seem like you're doing work!). Doing homework with a friend (e.g. testing each other) can work very well.

- **You need to give yourself rewards:** When you have achieved a certain number of objectives, why not reward yourself with a treat? This should help your motivation for the next series of tasks.

- **You need to take charge:** Revision is for *you* not your teacher. Remember it is what you learn that matters – and the more you learn, the better prepared you will be for the exam. So take charge!

Preparing for the examination

You should aim to complete most of your revision before the examinations actually start. Then you can spend any available time having a quick look at revision notes and relaxing between examinations. Here are some more tips:

- **Avoid overworking:** This can be counterproductive, and can make you confused, tired and increasingly worried. Some anxiety is useful – it can give you 'positive anticipation' for the exam. However, too much anxiety can diminish your performance. You can manage anxiety by careful planning of your workload, and by various relaxation techniques.

- **Adopt a positive attitude to your work and yourself:** Careful revision helps you to prepare for the exams, and builds confidence.

- **Find out where your strengths and weaknesses are:** Remember that examiners are interested in what you can do, rather that what you cannot do. This should be your attitude too. Identify where you have difficulties in understanding, and work on those topics especially (even though it may be tempting just to avoid the problem areas).

- **Monitor your revision progress:** You could check off topics in your notes, in the syllabus or in this book, when you are confident of your understanding. Then move on to a new topic.

- **Concentrate on revision tasks, rather than yourself:** Deal with difficult topics as problems to be solved, not as evidence of your inability! However, you should also give yourself rewards when you have achieved particular goals.

- **Know when and where the examination will be held:** Also, confirm in advance what paper (including what tier) you are taking, and what sort of questions you can expect. Remember what equipment (a calculator, for instance) to take to the exam.

IN THE EXAMINATION

Five things to remember

- **Read the general instructions, and the questions, carefully:** Many candidates lost marks simply because they have misunderstood what they have to do.

- **Keep your answers concise, accurate and relevant:** If you write a rambling, poorly organized answer you may lose marks and you will waste time. If possible, plan your answer (in your head) before you write it.

- **Allocate your time according to the marks available:** Even before the exam, you can calculate approximately how many minutes to spend for each mark available. For example, if the total number of marks in a one-hour exam is 60, then each mark is 'worth' just under a minute of your time (you need to allow time for reading the question). If you spend ten minutes answering a five-mark question in that exam, you will have problems completing the exam in time.

- **Answer the questions you know you can do first:** This will build your confidence for the remainder of the exam. Also, if you run out of time (which should not happen!), you can at least be sure that all questions that you could have answered easily have been answered.

- **You should already know what to expect!** Before the exam, it is important to make sure that you are already familiar with the types of question you are likely to face. These will vary between examining boards, syllabuses and papers, and will also depend on the tier you are studying.

Marking scheme

- **Mark allocation:** The written (terminal) examination in GCSE Physics will provide 75 per cent of your total assessment (the rest is coursework).

- **Mark distribution:** Within the written examinations, marks are distributed as follows:

  - 60 per cent of the total marks available are awarded for *knowledge, understanding* and *ability to apply information*. About one-third of these marks are based on your ability to recall information from memory. The other two-thirds will be based on information contained in the question.

  - The remaining 15 per cent of marks are given for *communication* and *evaluation*.

Types of questions

Structured questions

These are the most common form of question; some examination boards only use structured questions.

The basic format is to break the question down into sections (a), (b), (c), etc. Often, related parts of the question are grouped together, so that section (a) might be further broken down into subsections (i), (ii), (iii) and so on.

Often groups of questions follow on from each other; this makes it very important to get the first part right! It is usual for structured questions to become more demanding as you work through them. For instance, section (a) might ask for a basic description, (b) might be more searching, and (c) might ask about the applications of the chemical principles involved.

Mixed questions

Remember that examiners sometimes mix different themes in the same question. Here are some examples:

Question 1 (a), **(b)** – theme: periodic table.
(c) – theme: electronic structure of atoms.

Question 2 (a), **(b)** – theme: manufacture of aluminium.
(c) – theme: pollution.

The best way of approaching these questions is to be ready for a change in theme as you progress through the question. Avoid deciding that a question is all about just one theme, such as separation of mixtures; that way, you will not be fazed when part of the question suddenly switches to an entirely different theme.

There are 'warning signs' which you can look out for, such as the beginning of a new section (e.g. (b), (c)) – as you can see from the examples above, these changes will occur as you move from one section to the next. Also, changes usually occur towards the end of a question (if there are three sections then section (c) will often introduce a new theme).

Open-ended questions

These are only used with some examination boards and with certain papers, and are usually intended for higher tier candidates. Open-ended questions require an extended piece of writing (e.g. a paragraph). You need to plan your answer carefully: you must avoid 'going astray' or including irrelevant or 'rambling' material, which Examiners do not enjoy marking! It is important to practise the skills needed for answering these questions.

ABSOLUTE ZERO

This is the lowest possible temperature and is used as the zero of the kelvin **temperature scale**. It is at minus 273 °C.

When you heat up a material you give its particles more energy and they move faster. If you cool it down the particles move more slowly. You notice this as a change in temperature. If you cool the material enough its particles will almost come to a stop and you cannot take out any more energy to make it cooler. You have then reached absolute zero.

-◈- **Kinetic theory, Temperature, Temperature scale**

ACCELERATION

Acceleration is the rate of change of velocity.

The units will be m/s² because they are velocity units per second = m/s per s = m/s².

You can usually find acceleration from the formula:

$$\text{acceleration} = \frac{\text{change in velocity}}{\text{time taken}}$$

$$a = \frac{v - u}{t}$$

where a = acceleration, v = final velocity, u = original velocity, t = time in seconds.

If the answer to your calculation is a *negative* number it means that the body is slowing down rather than speeding up – it is *decelerating*. The acceleration is a **vector** because it takes place in a particular direction.

Worked example
A car moving at 10 m/s accelerates to 19 m/s in 12 s. What is its acceleration?

$$\text{acceleration} = \frac{v - u}{t}$$

$$= \frac{19 - 10}{12}$$

$$= \frac{9}{12} = 0.75 \text{ m/s}^2$$

CHECKPOINT

The driver of a train moving at 25 m/s brakes until the velocity becomes 10 m/s. If the braking takes 30 s what is the deceleration?

Remember: You can also find the acceleration from a **velocity/time graph** and you should remember that it will need a **force** to produce an acceleration.

-◈- **Force, Vector, Velocity**

ALPHA PARTICLES (α)

Alpha particles are emitted from the nucleus of some atoms in the process of **radioactive decay**.

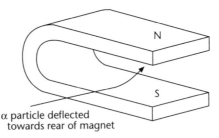

Effect of magnetic field

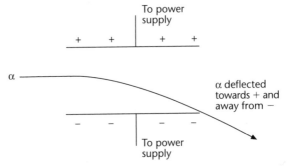

Effect of electric field

An α particle from a particular source will have a fixed range. It will be able to travel a few cm in air but will be stopped by a piece of paper. Because it is quite big and heavy compared with atoms, it will travel straight out from the source, producing **ions** by colliding with atoms and slowing down as it loses energy in each collision. It will make a very large number of ions in a short distance. It is made up of two **protons** and two **neutrons**, so it has the same **nucleus** as that of a helium atom, i.e. it has a **mass** of 4 and a **charge** of +2.

Its charge will cause it to be deflected away from the + and towards the − in an electric field and at right angles to a *very strong* magnetic field.

Whenever a **nuclide** decays by emitting an alpha particle the new atom left behind has a **proton number** that is 2 less than the original atom and a **nucleon number** that is 4 less than the original atom. See **Atom** for a definition of these numbers.

Remember that an alpha particle comes from the nucleus of the atom.
Examples:

$$^{241}_{95}\text{Am} \rightarrow {}^{237}_{93}\text{Np} + {}^{4}_{2}\alpha + \text{energy}$$

$$^{226}_{88}\text{Ra} \rightarrow {}^{222}_{86}\text{Rn} + {}^{4}_{2}\alpha + \text{energy}$$

1

Note that the totals of the nucleon and proton numbers will be the same on both sides of the equation. Americium-241 is only an alpha emitter, but others such as radium emit more than one type of radiation.

> *Remember: Because of the short penetrating range the dangers from alpha particles are small unless the source enters the body, but the source should still be handled at a distance using a handling tool and be stored in a properly labelled, lead-lined box.*

See **Radioactive decay** for uses.

✛ **Beta, Gamma, Half life, Nuclide, Radioactive decay**

ALTERNATING CURRENT (A.C.)

Alternating current will go first in one direction round a circuit and then in the opposite direction. This process is repeated many times per second – mains supply has a **frequency** of 50 Hz, so it changes direction 100 times per second.

This is done by keeping the *neutral* wire at the same voltage all the time, usually at earth potential, which we call 0 volts. The other, *live* wire, carries a voltage that increases to a maximum in one direction, falls to zero, increases to a maximum in the other direction, falls to zero and so on. When a complete circuit is connected to the supply the current will follow the changing voltage, first one way and then the other.

Measuring the current can be a problem, since its average value is zero (equal times at equal size in opposite directions) and its peak value only occurs at an instant and is smaller for the rest of the time. A common use of a.c. is to produce heat, which depends on the square of the current, and we therefore use a value called the *r.m.s.* (root mean square) value. This is the square root of the average of the square of the current and is the value of the d.c. that would have the same heating effect. It is about 0.7 times the *peak* value of the current. The voltages of alternating supplies are also stated as r.m.s. values.

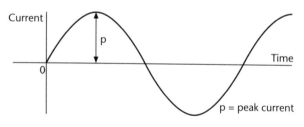

p = peak current

Domestic electricity supply is a.c. at 230 V r.m.s. with a frequency of 50 Hz (so the peak voltage will be almost 340 V!).

✛ **Current, Direct current**

ALTERNATIVE ENERGY SOURCES

These are sources of energy that do not use fossil fuels such as coal, natural gas and oil. They are usually renewable sources that will not run out if they are correctly managed.

✛ **Sources of energy**

ALTERNATOR

✛ **Dynamo**

AMMETER

An ammeter is a device for measuring electric **current** in **amperes** (amps). It is always placed in **series** with a **resistance** or a circuit component through which the current flowing is to be measured so that the current goes through the meter.

> *Remember: Ammeters should have a low resistance so that they do not reduce the flow of current.*

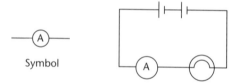

Symbol

Measuring the current through a bulb

✛ **Ampere, Voltmeter**

AMPERE

The ampere (amp) is the unit of electric **current**. An exact definition is not usually required at this level, but it is useful to think of it as the rate at which **charge** is being transferred round the circuit. Each **electron** carries such a tiny charge that one amp means 6×10^{18} (6 million million million!) electrons passing that point on the circuit each second.

The symbol for the amp is A.

✛ **Current**

AMPLITUDE

This is the maximum distance that a particle moves from its rest position during a vibration. The vibration is often the result of a **wave** passing and using the particle to carry its energy. It is measured in m or mm.

As a *transverse* wave (e.g. a water ripple) moves through a material the particles are vibrated at right angles to the direction of travel. In a *longitudinal* wave, (e.g. sound), the vibration will be backwards and forwards within the direction of travel of the wave. In both cases, the particle only vibrates about its rest position and does not move along with the wave. The amplitude is measured from the rest position to the extreme of its travel and *not* between the two extremes, which would be two amplitudes.

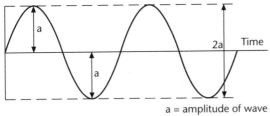

Displacement

a = amplitude of wave

A wave that is carrying more energy will have a greater amplitude of vibration as it passes. This will mean that a louder sound or a brighter ray of light will have a larger amplitude.

✦ *Frequency, Longitudinal wave, Transverse wave, Wave, Wavelength*

ANALOGUE

An analogue reading is one that can change smoothly and continuously without having to increase or decrease in steps. It can have any value within its range. Meters that use a needle on a scale give an analogue reading. A mechanical clock with a traditional face and fingers gives an analogue reading of the time. Most computers require a **digital** input. An analogue-to-digital converter is used to process the signals from sensors in experiments into a suitable form for a computer.

✦ *Digital*

AND GATE

✦ *Logic gates*

ANEROID BAROMETER

✦ *Barometer*

ANODE

Anode is the name given to a *positive electrode* either in an **electrolysis** cell or in an electrical cell (**battery**).

APPARENT DEPTH

The depth of a transparent material such as water or glass often appears to be less than it really is because the light is refracted at its surface.

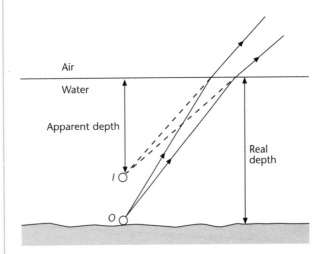

In the figure, as the light from an object *O* on the bottom of a pond passes through the surface it is refracted away from the normal so that it appears to come from the image *I*. The water will appear to be about three-quarters of its real depth. A stick or fishing rod that goes into the water will appear to bend as each part of it seems nearer to the surface than its real position. A swimming pool can seem deceptively shallow for the same reason. Glass will appear to be about two-thirds of its real depth.

✦ *Refraction*

ARCHIMEDES' PRINCIPLE

A body that is wholly or partly in a fluid (liquid or gas) has an upward force acting on it that is equal to the weight of the fluid that is displaced.

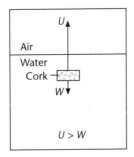

(a) Rises

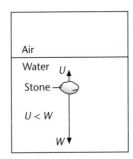

(b) Sinks

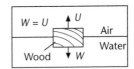

(c) Floats

The upward force is often called the 'upthrust'. If an object pushes out of the way a weight of fluid that is *greater than* its own weight, the upthrust is larger than the weight and the object moves *upwards*. This would be the case for a cork under water or a balloon filled with hydrogen or hot air.

In many cases the weight displaced is *less than* the weight of the object and the upthrust is therefore less than the weight. The object then moves *downwards*. This would be the case for a stone in water.

In a few cases the weight of fluid displaced is *equal to* the weight of the object so that there is no overall force up or down and the object *floats* or *hovers* at the same level. A submarine can do this by adjusting the amount of air and water in its ballast tanks. A ship will float when its weight is equal to the upthrust, so it settles into the water until it displaces its own weight of water. When the ship takes on cargo it gets heavier and settles further into the water to restore the balance by displacing more water.

-+- *Force*

ATMOSPHERIC PRESSURE

-+- *Barometer, Pressure*

ATOM

An atom is the smallest possible part of an **element**. There are ninety-two naturally occurring elements and another fifteen or so that can be produced with the help of a nuclear reactor. Each atom will consist of a central nucleus surrounded by **electrons** in energy levels (sometimes called 'shells' or 'orbits'). The nucleus is made up of **protons** and **neutrons**. Atoms are extremely small and even a speck of dust will contain huge numbers of them.

Since atoms are so small we use a new mass unit for them, so that one unit of mass is one-twelfth of the mass of an atom of the most common **isotope** of carbon. Each **proton** has a **mass** of one unit and carries a **charge** of $+1$. The *neutrons* have a mass of one unit and are neutral (so they carry no charge at all). The **electrons** are very tiny and their mass is so small that we often count it as zero, but each one carries a charge of -1.

Atomic particles and their properties		
Particle	*Mass*	*Charge*
Proton	1	$+1$
Neutron	1	0
Electron	0 (almost)	-1

The electrons around the outside of the atom are the parts that join atoms to make compounds and therefore decide the *chemistry* of the atom. Since atoms are neutral, each one has to contain the same number of electrons as there are protons in the nucleus. The number of protons in the nucleus will therefore decide the element to which the atom belongs.

Proton number (atomic number)

The *proton number* is the number of protons in the nucleus of an atom. All atoms of the same element will have the same number of protons. This number is written below, and to the left of, the symbol for the atom: $_6C$, $_{92}U$, $_1H$, $_8O$, $_{38}Sr$.

For example, all atoms of carbon will have six protons in their nucleus. They will also have six electrons in 'energy levels' outside the nucleus, which decide the chemistry of carbon.

Uranium has a proton number of 92, so every atom of uranium has 92 protons in the nucleus and 92 electrons in energy levels.

Nucleon number (atomic mass number)

The *nucleon number* is the *total* number of protons and neutrons in the nucleus of the atom. Sometimes the number of neutrons can vary in atoms of the same element, and this produces **isotopes**. The nucleon number is always written above and to the left of the symbol of the atom: ^{12}C, ^{238}U, ^1H, ^{16}O, ^{90}Sr. Subtracting the proton number from the nucleon number will tell you the number of neutrons in the nucleus.

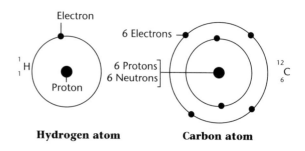

Hydrogen atom   **Carbon atom**

-+- *Electron, Element, Ion, Isotope, Neutron, Nuclide, Proton, Radioactive decay*

ATOMIC MASS NUMBER (NUCLEON NUMBER)

This is the total number of **protons** and **neutrons** in the nucleus of an atom.

-+- *Atom, Neutron, Proton*

ATOMIC NUMBER
(PROTON NUMBER)

This is the number of **protons** in the nucleus of an **atom**. The atomic number will decide which element the atom belongs to and where it fits in the periodic table.

Atom, Proton

BAROMETER

A barometer is used for measuring the **pressure** of the atmosphere.

There are several types, but the *aneroid barometer* is the one in most common use, now that people are more concerned about the danger of mercury vapour from the older liquid-in-glass types. The mercury types are also more difficult to read and easier to damage.

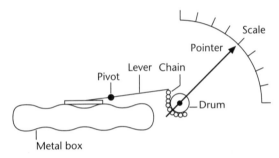

Aneroid barometer

The metal box is sealed and contains air at less than atmospheric pressure, so that it is slightly squashed. It has corrugated sides so that it can change shape slightly. If the pressure outside increases, the box is flattened a little more and the lever connected to the box is moved. The lever, chain and drum magnify the movement, so that a small change in pressure moves the pointer a long way on the scale. When the pressure falls again the movements are reversed. The scale is marked in millibars (1 bar is a pressure of 100,000 Pa) and these are the units used by weather forecasters.

Remember: The barometer only measures atmospheric pressure – it does NOT measure weather! It will give an indication of what might happen to the weather, as a falling reading usually means wetter weather, a steady reading means the weather staying the same and a rising reading means dryer weather. Clear skies from a high pressure are cold in winter and hot in summer. Low pressures give more cloud which is warmer in the winter but cooler in the summer. On a weather map places that have the same pressure are joined together with lines called 'isobars', and the pressure is written next to the isobar in millibars.

If we use a barometer we find that the average atmospheric pressure at sea level is 760 mm of mercury. This pressure is called a 'standard atmosphere' (See **Pressure.**) As you move up away from the surface of the Earth the air molecules are more spaced out and the pressure is lower. Climbers on Everest will need oxygen cylinders because the air's density is low. If you travel in a jet airliner it will have a 'pressurized' cabin that has an atmosphere at a normal pressure even when the pressure outside is very low because of the altitude of the aircraft. It is possible to use a barometer as an altimeter by recalibrating its scale.

⟡ **Pressure**

BATTERY

A battery is made up of **cells** connected in **series**. The **electromotive force** of the battery is the sum of the e.m.f. of all the cells. A car battery contains six 2 V cells in series.

6 V

Remember: Voltages in series add.

⟡ **Cell, Electromotive force**

BETA PARTICLES (β)

Beta particles are emitted from the nucleus of some atoms as part of the process of **radioactive decay**.

A beta particle from a particular source will have a fixed *maximum* range. It will penetrate paper and may travel several metres in air. It will be stopped by a few mm of aluminium. It is very small and light compared with an **atom** but will be moving very fast. It will travel out from the source, creating **ions** as it collides with atoms and knocks electrons off them. The particle will be slowed by each collision and, because of its very small mass, will also probably be deflected. The trail of ionisation left by the beta particle is longer than that for an **alpha particle**, but the ions are more spaced out and the track keeps changing direction instead of being short and straight.

The particle is actually a fast-moving electron, so its charge is −1 and it has a very small mass. It will be deflected away from the − and towards the + in an electric field. In a magnetic field it will obey **Fleming's left-hand rule**, but you will need to remember that the conventional current is in the opposite direction to the electron flow.

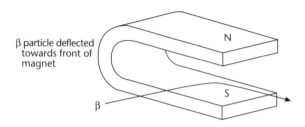

β particle deflected towards front of magnet

Effect of magnectic field

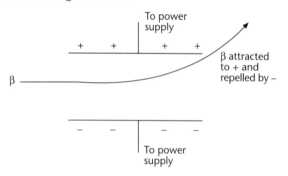

β attracted to + and repelled by –

Effect of electric field

It is much easier to deflect a beta particle than an alpha particle, and a fairly strong magnetic field can even trap the particles so that they are going round in complete circles.

Whenever a **nuclide** decays by emitting a beta particle the nucleon number remains the same but the new atom formed has a proton number that is 1 *larger*. (See **Atom** for a definition of these numbers.) Remember that the beta particle comes from the nucleus of the atom.

For example, the nuclide strontium-90, $^{90}_{38}$Sr decays as follows:

$$^{90}_{38}\text{Sr} \rightarrow {}^{90}_{39}\text{Y} + {}^{0}_{-1}\beta + \text{energy}$$

Note that the totals of the nucleon and proton numbers will be the same on both sides of the equation.

*Remember: Beta particles are quite penetrating and, as a short-term effect, can produce burns and sores on the skin. Their more dangerous effects will be in the body, where in the long term their damage to cells will cause cancers. 'Fall-out' from nuclear weapons contains sources like this that would be absorbed into the bones and have a long **half life**, causing serious illness. It is important to avoid close contact with these sources, and they should always be handled at a distance with the proper handling tool. They must be kept in the correct, labelled, lead-lined box.*

See **Radioactive decay** for uses.

✛ **Alpha particles, Atom, Gamma radiation, Half life, Radioactive decay**

BIG BANG THEORY

This is one theory about how and when the **universe** was created. Astronomers believe that the universe is expanding continuously as though it all came from one place. If we go back in time (by about 15 billion years), all of the matter in the universe would have been in one place. It then exploded outwards in what is now called the 'Big Bang'.

We know that all of the other galaxies are moving away from us because of the **red shift** in the light that we receive from them.

We can detect the remains of the radiation from the Big Bang as it spreads out in space.

This does not explain how all of the matter got into one place at the start. One theory is that gravity will eventually stop the universe expanding and pull it all back together again in a Big Crunch! The whole thing could then begin again with another Big Bang.

There are other theories about how the universe was created but, at the moment, we think this is the most likely. For example, one says that matter is always being created in the new space that is being created as the universe expands.

✛ **Galaxy, Red shift, Universe**

BIMETALLIC STRIP

A bimetallic strip consists of two strips of two metals bonded together. When the **temperature** of the strip changes one of the metals expands or contracts more than the other, causing the strip to bend so that the longer metal is on the outside of the curve. The alloy 'invar' is often used as one of the strips because it does not expand much (see **Expansion of solids**).

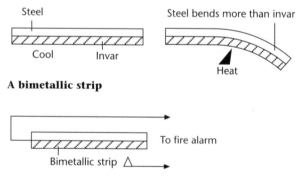

A bimetallic strip

A fire-alarm circuit

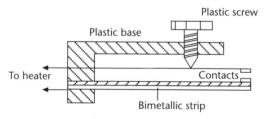

An adjustable thermostat

There are a number of uses for the bimetallic strip, including simple fire-alarm circuits and adjustable thermostats. The bimetallic strip in the thermostat will bend on cooling so that the contacts touch and turn on the heater. When the temperature reaches a pre-set value the strip has bent enough the other way for the contacts to separate and turn off the heater. The temperature at which this happens can be adjusted by changing the distance to the other contact by turning the adjustment screw.

Expansion of solids

BIOMASS

Biomass is an *energy* source that starts as vegetation, all of which has used energy from the Sun in a photosynthesis reaction to build up more complex chemicals. Using biomass involves using the sunlight to grow plants with a high *energy* content, usually in the form of sugars. The energy is then extracted from the chemicals produced by the plants. In Brazil, sugar is extracted from plants and fermented to produce alcohol. This can then be used instead of petrol to power car engines.

Sources of energy

BOILING POINT

Boiling point is the temperature at which a liquid boils under a pressure of 1 atmosphere (760 mm of mercury pressure).

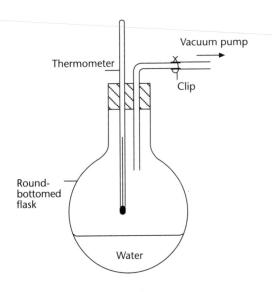

A liquid will boil when the pressure of its *vapour* is equal to the pressure in the space above its surface. If air is pumped out of a round-bottomed flask, gentle heating will show that the water now boils at a lower temperature than the expected 100 °C.

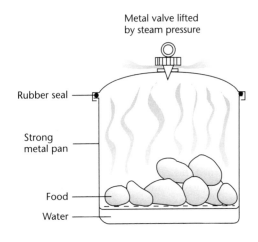

A pressure cooker

If the pressure above the surface is increased, the boiling point will be correspondingly higher. A pressure cooker works on this principle, the steam being able to escape through the valve only when the pressure has been raised. The boiling point of the water, and therefore the temperature inside the cooker, is higher and the food cooks more quickly.

The boiling point of a liquid is also raised by dissolved substances. The boiling point of water, for instance, is raised by dissolving salt in it. (There is scope for an investigation here! Does the quantity of salt matter? If so, what concentration produces what effect?) Chemists use boiling points as a check for purity.

Freezing point, Kinetic theory, Vapour

BOYLE'S LAW

If a fixed mass of gas is kept at a constant temperature, its pressure is inversely proportional to the volume.

If the mass of gas is fixed, there will always be the same number of molecules in the container. The molecules will always be moving at the same velocity because the temperature is constant, and they will collide with the walls of the container, causing a pressure (see **Kinetic theory**). If the volume is halved, the molecules will hit the walls twice as often and the pressure will double. This inverse proportion always happens, so that whatever you divide the volume by, the pressure is multiplied by that number. This means that if you can measure the pressure and volume and multiply the answers together the answer always stays the same until the mass or temperature of the gas is changed.

$$\text{pressure} \times \text{volume} = \text{constant}$$

$$P_1 \times V_1 = P_2 \times V_2$$

Where P_1 = first pressure, V_1 = first volume, P_2 = second pressure, V_2 = second volume.

9

Worked example

A cylinder contains 20 litres of gas at 25 atmos. pressure. What volume of gas would there be at 1 atmos. pressure?

$$P_1V_1 = P_2V_2$$

$$25 \times 20 = 1 \times V_2$$

$$500 = V_2$$

new volume = 500 litres

If you owned this cylinder you would probably get only 480 litres of gas from it, as the gas would stop coming out when the pressure inside was the same as that outside, i.e. 1 atmos.

CHECKPOINT

A small bubble is released at the bottom of a lake that is 20 m deep. If the volume of the bubble is 2 cm³, what will it become when it reaches the surface of the lake? (10 m of water has the same pressure as 1 atmosphere.)

An experiment to show Boyle's law can be done with the special apparatus shown in the figure. Air is pumped into the air reservoir with a bicycle pump. The pressure can be read on the pressure gauge and the volume can be read from the scale at the side of the tube. Some air is then released from the valve to reduce the pressure and, after waiting a short time to make sure the temperature is the same as before, a new pair of readings is noted. This is repeated several times.

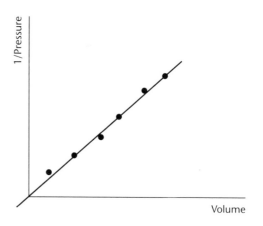

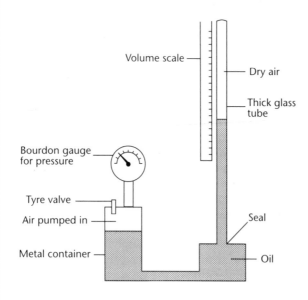

The results are noted in a table with the following headings:

Pressure	Volume	1/Pressure

Plotting a graph of pressure against volume will produce a smooth curve, which shows that there is connection between the two quantities. Only a straight line shows the exact connection. You can get this by plotting 1/pressure against volume, which proves that the law is true.

Check your own syllabus carefully to find if you need to know all this work for your examinations. Some syllabuses only require you to be able to *describe* how pressure and volume affect each other.

 Charles' law, General gas equation, Kinetic theory, Pressure law

BROWNIAN MOTION

This is the random motion of a particle caused by molecules of a gas or a liquid colliding with it.

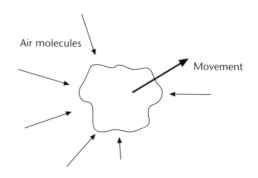

Dust particle in air

A small particle such as dust or smoke will have a random motion even in still air, as it is in collision with the molecules of the air. The motion of the molecules is random, and therefore at any given time, there are more molecules hitting one side of the dust than the other, so that it is pushed in one direction. A short time later the force may be in another direction. This will not occur with large particles, because the total of the many forces in each direction will then become the same.

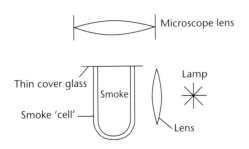

We can observe this movement by looking at smoke through a microscope. The smoke in the cell is seen as tiny bright dots of reflected light that are in continuous random motion. A similar experiment will show the same motion when you look at small crystals or poster paints in a clear liquid.

> Remember: The motion shows that particles in gases and liquids are moving as described in the kinetic theory.

Kinetic theory

CAMERA

The simple camera is a light-proof box with a **lens** at the front that can produce an image on the film. The image will be inverted and smaller than the object. This happens for only a very short time – about 1/100 s – when a hole in the shutter is opened to expose the film. The film is usually in a cassette so that it can be wound on after each picture, ready for the next.

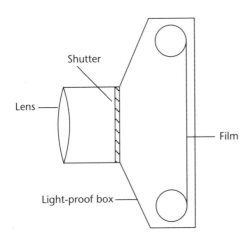

A *ray diagram* is shown under **Lens**.

A simple camera can be improved in a lot of ways. For instance, you can:

● Fit the lens in a screw mount so that it can be moved backwards and forwards. This enables objects at different distances to be focused sharply.

● Have a shutter that can open for different times so that you can take pictures of fast-moving objects without them being blurred.

● Have a diaphragm behind the lens. This is a way of changing the size of the hole that the light goes through so that the same amount of light can reach the film on a bright or dark day. The size of the hole is called the aperture.

● Have a built-in meter to measure the light. This will make sure that you set the shutter and aperture to get the correct light to the film.

 Lens

CAPACITOR

A capacitor is a device that is used to store electric **charge.**

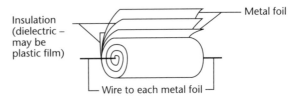

Simple metal foil capacitor

Symbol   **Electrolytic capacitor**

As charge is put into the capacitor, the **potential difference** (voltage) across it rises.

There are several main uses for a capacitor, each based on different properties:

● As a storage capacitor. (See **Half-wave rectifier** and **Full-wave rectifier**.) The charge stored by a capacitor is not especially large but is perfectly adequate for supplying the small currents used in electronic circuits.

● As a block or filter for removing d.c., which cannot pass through a capacitor, because there is no complete circuit; the current stops when the capacitor is fully charged. The a.c. can pass by repeatedly charging the capacitor, first in one direction and then in the other. Putting a capacitor into a circuit can therefore separate d.c. from a.c.

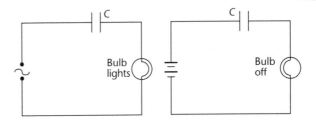

Capacitor *C* in a.c. and d.c. circuits

● As part of a timing circuit. If the capacitor is charged through a resistor it will take a short time for the p.d. on the capacitor to reach its maximum value. A larger capacitor or a larger resistance will make the time longer. The potential divider shown could be connected to the input (base) of a **transistor switch** circuit. The transistor would come on when V_{out} reached the necessary 0.6 V.

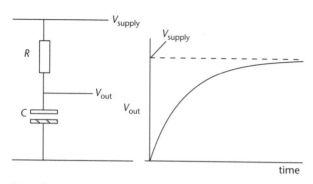

Charging a capacitor

The ability of the capacitor to store charge is called its *capacitance* and is measured in *farads*.

A capacitor will have a capacitance of 1 farad if a p.d. of 1 volt is produced by putting 1 coulomb of charge on it.

In real circuits the farad is too large and we work in millionths of farads, called microfarads (μF). The definition also tells us that:

$$\text{charge stored} = \text{capacitance} \times \text{p.d.}$$
$$Q = CV$$

Worked example

How much charge is stored on a 500 μF capacitor when it is charged so that the p.d. across it is 15 V?

$$\text{charge stored} = \text{capacitance} \times \text{p.d.}$$
$$= \frac{500 \times 15}{1,000,000}$$
$$= 0.0075 \, C$$

Each capacitor will be marked with a maximum working voltage and it is important to make sure that you do not exceed this value – it is easy to make this mistake by forgetting that the peak value of an alternating voltage is 1.4 times the r.m.s. value that you measure or calculate and that the capacitor will usually charge up to the peak value! (See **Alternating current**.) Some capacitors, especially the 'electrolytic' types, need to be connected the correct way round in a circuit, and their leads will be marked $+$ or $-$.

⊶ **Charge, Rectifier**

CATHODE

The cathode is the name given to a *negative* electrode either in an **electrolysis** cell or an electrical cell (**battery**).

CATHODE RAY OSCILLOSCOPE (C.R.O.)

⊶ **Oscilloscope**

CELL

An electric cell produces **current** from a chemical reaction inside it. It is often called a **battery**. A battery is really two or more cells joined in **series**. There are two main types:

Primary cells

Primary cells cannot be recharged. They produce a current by a chemical reaction that cannot be reversed. When one of the chemicals is used up the reaction will stop and the cell produces no more current, e.g. simple cell, dry cell.

A *simple cell* will be made from two different metals put into a dilute acid (strips of copper and magnesium pushed into a lemon will run a small motor!). The voltage will be bigger if the metals are further apart in the activity series.

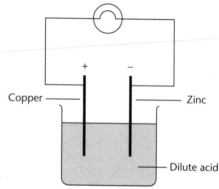

A simple cell

The less reactive metal (in this case copper) will be the $+$ and the more reactive (in this case magnesium) will be the $-$. The cell produces current to light the bulb for only a short time before the copper plate is coated with small bubbles of hydrogen, which act as an insulator. This is called *polarization*.

Polarization can be stopped by adding an oxidizing agent to turn the hydrogen into water. Another problem is *local action*, where the more reactive metal (the zinc) is rapidly dissolved into the acid by small impurities in the metal forming small cells. Disadvantages include an acid that is easy to spill and a low voltage produced for a short time.

A *dry cell* is the common sort of cell that you would buy to put into a torch. It will produce about 1.5 V.

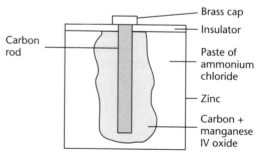

A dry cell

In this case the carbon rod is the + and the zinc is the −. The cell may leak when the zinc is used up and the paste leaks through the holes – this can be prevented by an outer coating of plastic on the sides. There are no liquids to spill and the manganese oxide gets rid of the hydrogen, so the cell works for quite a long time.

Secondary cells

Secondary cells can be recharged. The chemical reaction that produces the current can be reversed by sending a charging current through the cell in the opposite direction, e.g. lead–acid cell (as in a car battery), nicad.

A *lead–acid cell*. The cell is first charged by sending a current through it. The + plate becomes coated with a layer of brown lead oxide and the − plate is coated with grey lead. The cell produces about 2 V and has a low resistance, so it can give out a large current (which the dry cell cannot!). When the cell is discharged it turns the coatings into lead sulphate.

The cell gives off hydrogen and oxygen by splitting up water when it is being charged and this may need to be replaced by adding pure water.

The symbol for one cell

CHECKPOINT

What is the difference between a primary and a secondary cell?

Name an example of each.

Give some advantages of using a secondary cell to power a personal stereo.

-❖- **Battery, Current, Electromotive force**

CELSIUS

This is the **temperature scale** that was invented by Anders Celsius and used to be called 'centigrade'. The temperature at which pure water freezes is given the value 0 °C. The temperature at which pure water boils under 1 atmosphere pressure is 100 °C. The temperature between these 'fixed points' is then divided into 100 'degrees'.

-❖- **Temperature, Temperature scale**

CENTRE OF MASS

A body will often behave as though all of its **mass** is concentrated at a point, called its centre of mass.

It will also behave as if all its **weight** is concentrated at the same place and, for that reason, this point is also called the *centre of gravity*.

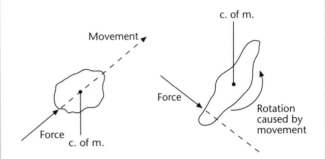

> *Remember: If a force acting on the body points at the centre of mass, it will accelerate along the line of force. If the force points to one side of the centre of mass, it will cause the object to rotate because a MOMENT is produced. If there is no pivot, the body will also move forward.*

-❖- **Equilibrium, Mass, Moment, Weight**

CENTRIPETAL FORCE

This is the **force** acting on a body that is moving in a circle or along a curve. It acts *towards* the centre of the circle.

If a body moves at a constant speed in a circle, its direction is changing all the time. This means that the **velocity**, which is a **vector**, is also changing all the time. In order to change the velocity, **Newton's first law** tells us that there must be a force.

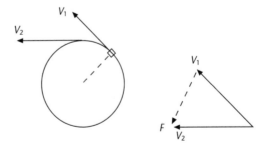

The figure shows a body moving in a circle. At that instant its velocity is in the direction shown by V_1. A short time later the velocity has become V_2. There must be a force F acting towards the centre of the circle to change the velocity. The force needed is larger if the mass is larger or if the velocity is larger.

The force gets smaller if the radius of the circle gets larger. For example:

● The Moon is attracted by the Earth and this provides the centripetal force that keeps it in orbit. All satellites need a centripetal force, which is provided by gravity.

● A car going fast around a bend produces the centripetal force by the friction of its tyres on the road. If it goes into the bend too fast or if the road is icy, the tyres cannot produce a force that is big enough and the car will go on in its original direction – hitting the outside of the bend!

● If wet clothes are rotated quickly in a spin-dryer the fabric cannot produce a centripetal force large enough to keep the water going quickly in a small circle. The water moves in a line as straight as it can – to the outside of the drum, through the holes to the outer drum, where it is collected.

-⊹- *Force, Velocity, Vector*

CHAIN REACTION

This is a series of reactions in which each one triggers the next. It usually refers to the *fission* of large *atoms*. This is caused by the nucleus being hit by a *neutron* – each fission produces more neutrons to hit more nuclei and keep the chain reaction going.

-⊹- *Fission, Nuclear reactor*

CHANGE OF STATE

Materials can be either *solid, liquid* or *gas*. These are called *states of matter*.

When a kettle is boiling, the water stays at 100 °C and the *energy* being put in is used to change the liquid water into steam, where the particles are a lot further apart. In a similar way, ice will melt at 0 °C, the heat energy supplied being used to break down the structure of the solid and make the particles free to flow.

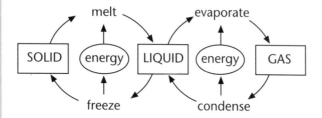

Energy and changing state

Energy is needed to make a material melt, and the same energy is released when the material freezes. In a similar way, energy is needed to vaporize a liquid and the same energy is released when the vapour condenses. This energy is called *latent heat*. The latent heat to vaporize a liquid will be a lot greater than the latent heat that melted the solid. Different materials will change from one **state** to another at different temperatures.

> Remember: A change of state will happen at a constant temperature that is special for that material. The temperature is the **freezing point** or **boiling point** of the material.

SOLID   LIQUID

• Regular arrangement
• Strong forces between particles
• Particles only vibrate

• Irregular arrangement
• Strong forces between particles
• Particles move more than in solids

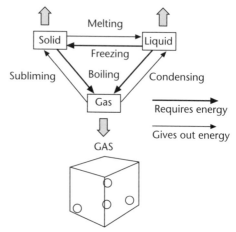

GAS

• Irregular arrangement
• No forces (or only weak forces) between particles
• Rapid movement of particles but no pattern to the movement

Changing state

A refrigerator is cooled by a liquid that is evaporated in metal pipes in the freezer compartment. The heat energy needed to do this is taken from inside the fridge. The compressor then turns the vapour back into a liquid inside the radiator on the back of the fridge and the heat energy is released into the surrounding air.

Liquids that evaporate quickly feel cold if they are spilled on your hand. This is because they obtain the heat energy needed to evaporate from your hand. A burn caused by steam will often be much worse than a burn from hot water because of the large amount of extra heat that is released as the steam turns back into hot water.

Do not forget that steam is a colourless gas and should not be confused with the clouds of small water droplets that we often call 'steam' in everyday English. Look for steam close to the spout of a boiling kettle but DO NOT TOUCH!

Water can be used as an example of these different states. As ice it is solid, as water it is liquid and as steam it is gas.

> Remember: Look carefully at the diagram to see how **kinetic theory** says the particles are arranged in each of the states.

CHECKPOINT

Complete the labels on the diagram.

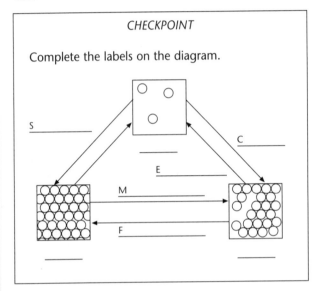

S _____

C _____

E _____

M _____

F _____

-|- **Kinetic theory, State**

CHARGE

Electric charge is a property of **electrons**, which carry a tiny negative charge, and **protons**, which carry a tiny positive charge (see **Static electricity**). As the electrons move around an electric circuit they make an electric **current** because each one carries a tiny negative electric charge.

The electric charge transferred through any point in a circuit by an electric current is given by:

$$\text{charge transferred} = \text{current} \times \text{time}$$

$$Q = It$$

The charge will be in **coulombs** (C) if the current is in amps and the time is in seconds. This means that 1 coulomb of charge is transferred around each part of a circuit when 1 A flows in it for 1 s.

Worked example
How much charge flows around the circuit when a car battery produces 7 A for 2 hrs?

$$\text{charge} = \text{current} \times \text{time}$$
$$= 7 \times 2 \times 60 \times 60$$
$$= 50{,}400 \text{ C}$$

CHECKPOINT

Write down the formula that connects charge, current and time. How much charge flows through a resistor if it carries a current of 0.25 A for 2 minutes?

> Remember: Each moving electron carries a VERY TINY CHARGE, so that huge numbers are moving along a wire to transfer a coulomb of electricity from one place to another.

-|- **Capacitor, Current, Static electricity**

CHARLES' LAW

If a fixed **mass** of gas is kept at a constant **pressure**, its volume is directly proportional to its absolute temperature.

If the mass of gas is fixed, there will always be the same number of molecules in the container. If the temperature rises, the molecules go faster and hit the walls harder (see **Kinetic theory**). This raises the pressure slightly and the walls are pushed out a little until the number of collisions (and the pressure) falls to what it was before. The increase in temperature has therefore produced an expansion. The volume will be proportional to the temperature provided that the kelvin temperature scale is used.

$$\frac{\text{volume}}{\text{temperature}} = \text{constant}$$

Or

$$\frac{V_1}{T_1} = \frac{V_2}{T_2}$$

Where V_1 = first volume, T_1 = first temperature, V_2 = second volume, T_2 = second temperature.

Temperatures *must* be in kelvin. Add 273 to the temperature in °C.

Worked example
A gas occupies 10 litres at 27 °C. What will its volume become at 127 °C and the same pressure?

The pressure is constant, so we may use Charles' law.

$$T_1 = 27 + 273 = 300 \text{ K}$$
$$T_2 = 127 + 273 = 400 \text{ K}$$
$$\frac{V_1}{T_1} = \frac{V_2}{T_2}$$
$$\frac{10}{300} = \frac{V_2}{400}$$
$$V_2 = \frac{400 \times 10}{300}$$
$$= 13.3 \text{ litres}$$

> **CHECKPOINT**
>
> 500 ml of gas at 100 °C is collected from an experiment. What will the volume become at 20 °C and the same pressure?

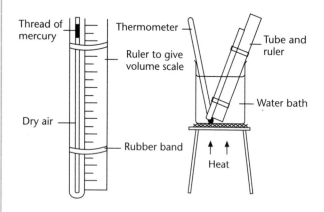

An experiment to show Charles' law can be done with the apparatus shown in the figure. The dry gas is trapped by a thread of mercury in a capillary tube, which has been sealed at one end. The gas can expand freely by moving the mercury, and its pressure is always close to atmospheric pressure, since the tube is open to the air at one end. The tube is the same diameter all the way along, so the volume of gas is found as the length of the air on the ruler. The temperature of the water bath is kept constant for a few minutes and the volume of the gas read on the ruler. This is noted together with the temperature reading on the thermometer. The temperature of the bath is then changed and the reading repeated until a set of values has been obtained.

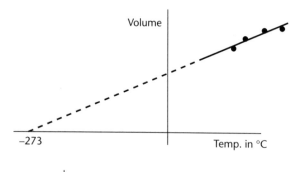

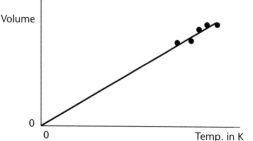

A graph is then plotted of volume on the vertical axis and temperature on the horizontal axis. This

turns out to be a straight line that can be extrapolated back to cut the temperature axis at −273 °C, i.e. the volume would become zero at −273 °C if the gas carried on behaving in the same way. The straight line shows that the law is true but that the temperature *must* start from −273 °C, which we call *absolute zero*. The temperatures will therefore be in kelvin.

> *Remember: Check your own syllabus carefully to find out if you need to know all of this work for your examination. Some syllabuses only need you to be able to DESCRIBE how temperature and volume are related.*

✦ *Boyle's law, General gas equation, Kinetic theory, Pressure law*

CHEMICAL ENERGY

Chemical energy is contained in the bonds between atoms in **compounds**. If the atoms become bonded in a different way, or different compounds are formed, the bonding energy required may be less and heat will be released – an *exothermic* reaction. The different fuels that we burn may give out different quantities of heat, but they all release some of their chemical energy as heat energy when they burn. Some reactions take heat energy in because the new compounds require more bonding energy. These reactions are called *endothermic* reactions.

✦ *Energy*

CIRCUIT

An electric circuit will have a power supply (a battery, generator or the 'mains') and something that is given energy from the supply (bulb, heater, motor).

Current will flow only when there is a complete circuit from one side of the supply all the way around to the other side. Switches turn the current off by making a gap in the circuit.

Each piece of the circuit will have its own **circuit symbol** to make the circuit easier to draw.

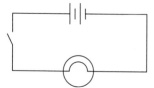

The circuit is complete only when the switch is closed.

> *Remember: Current is pushed around the circuit by the voltage produced by the supply. The current moves around the circuit but the voltage stays where it is!*

-✦- *Cell, Current*

CIRCUIT SYMBOLS

-✦- *Appendix Three*

COLOUR

The different frequencies of light have different effects on the retina of the eye, so we see them as different colours.

-✦- *Electromagnetic spectrum, Primary colours, Spectrum*

COMPOUND

A compound is a substance that contains atoms of more than one element bonded together chemically. This bonding makes the chemistry of the compound quite different from that of a simple mixture of the same atoms. The particles in a compound may be either molecules or ions. For example, water is a compound. It is made up of molecules that each contain two hydrogen atoms and an oxygen atom bonded together. Sodium chloride is made up of sodium ions and chlorine ions. These ions have opposite electric charges and attract each other strongly, forming the bond.

-✦- *Element, Mixture*

COMPRESSION

A pressing or squeezing *force.*

-✦- *Hooke's law, Tension*

CONDUCTION OF HEAT

Conduction of heat can occur in two ways:

1. The heat at one end of a piece of material causes the particles there to vibrate rapidly and the temperature rises (***kinetic theory***). This vibration is passed into the next layer of particles as they collide. The heat energy is passed from one layer to another until it reaches the colder end. This does happen, but it is not a very efficient process and it may take a large temperature difference to drive much heat through the material. If the particles are large ***molecules*** the result is likely to be a poor conductor or an ***insulator*** (e.g. plastics).

2. The heat is carried through a metal by 'free' ***electrons*** in the metal structure. All metals are good conductors.

Some materials will be better conductors than others – they have a better *conductivity*. The figure shows a simple method of comparing the conductivity along rods with the same dimensions. The wax will melt first on the best conductor, and the drawing pin will fall off.

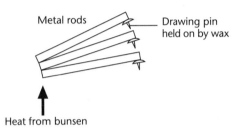

Liquids are often poor conductors, and heat travels through them by ***convection***. If the liquid is heated at the top, some heat is conducted down but not as quickly as in a solid. This is shown in the following figure. The water at the top is boiling but the ice remains frozen at the bottom. This property is used in hot-water tanks, which are filled with hot water from the top downwards so that you do not need to heat as much water.

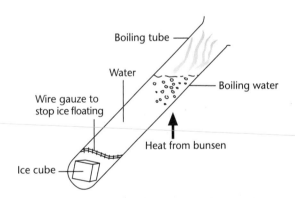

Gases are very poor conductors because the molecules are too far apart, and heat is transferred through them by convection. Trapping pockets or layers of air can make a good insulator. Double glazing works by trapping a layer of air between two layers of glass. See other examples under ***Insulator***.

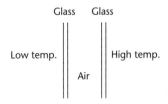

Double glazing: twice as much glass and an extra layer of a poor conductor

There are many uses for good conductors, such as in pans, boiler pipes and the walls of car engines. They are all made from metal.

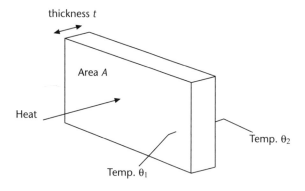

thickness t

Area A

Heat

Temp. θ_2

Temp. θ_1

The amount of heat through the block of material in the figure above will depend on the material from which it is made. It will also depend on:

1. The area A. For example, more heat can be lost through a large window than a small one.

2. The thickness t. More heat will be lost through a thinner material than through a thicker one. A thick layer of fibreglass makes a better insulator than a thin one!

3. The temperature difference across the material $(\theta_1 - \theta_2)$. A greater temperature difference results in more heat passing through.

CHECKPOINT

Which materials are always good conductors? Explain how heat energy is transferred in these good conductors.

Name two practical uses for good conductors and two for good insulators.

-·- **Conductor, Convection, Insulator, Infrared radiation, U-values, Vacuum flask**

CONDUCTOR

A conductor is a material that allows heat or electricity to flow through it easily.

All metals are good conductors of both heat and electricity. They have a lot of electrons, which can move about quickly and easily within the structure of the metal, carrying both electric **charge** and some **kinetic energy**. The moving charges are electric **current** and the kinetic energy is the heat being carried through the material. The only non-metal that is a conductor of electricity is carbon, which has a fairly high resistivity and is used to make **resistors**. The best electrical conductor is copper, and this is

used for household cables. However, in very large cables it can be both heavy and expensive and aluminium may be used instead.

-·- **Conduction of heat, Current, Insulator**

CONSERVATION OF ENERGY

Energy is never created or destroyed; the total amount remains constant. Energy sometimes appears to have been lost, but it will have been transferred into another system or converted into a form that you do not want.

Remember: This law can save you a lot of time and calculation; e.g. as a high jumper crosses the bar she will have gravitational potential energy, which you can calculate. When she reaches the ground the same amount of energy will have become kinetic energy – so you do not need to work it out again.

-·- **Energy**

CONSERVATION OF MOMENTUM

When two or more bodies collide, the total **momentum** *after* the collision is the same as the total momentum *before* the collision.

This is true in all cases from snooker balls colliding to explosions, in which the total momentum of all the pieces after the explosion is the same as the total before. At this level of examination some of these cases are too difficult to analyse, and set problems are restricted to those events where all the bodies stay in a straight line. Problems are also much easier to solve if all the objects are stuck together either before or after the event, so that the total momentum is easy to calculate.

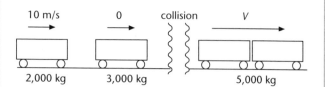

10 m/s   0   collision V

2,000 kg   3,000 kg   5,000 kg

Worked example

A railway truck (see figure) has a mass of 2,000 kg and is moving along the track at 10 m/s when it hits a stationary truck of mass 3,000 kg. The two automatically couple together and continue to move along the track.

1. How much momentum do the two trucks have before the collision?

> momentum of truck 1 = mass × velocity
> $$= 2,000 \times 10$$
> $$= 20,000 \text{ kg m/s}$$

> momentum of truck 2 = mass × velocity
> $$= 3,000 \times 0$$
> $$= 0 \text{ kg m/s}$$

> total momentum before collision = 20,000 kg m/s

2. How much momentum do the trucks have after the collision?

> 20,000 kg m/s (conservation of momentum)

3. What is the **velocity** of the trucks after the collision? Note that both trucks are now coupled together.

> total mass of trucks = 2,000 + 3,000 = 5,000 kg

> momentum of trucks = mass × velocity

> $$20,000 = 5,000 \times \text{velocity}$$
> $$\text{velocity} = \frac{20,000}{5,000}$$
> $$= 4 \text{ m/s}$$

> *Remember: A sketch diagram often helps to make the problem clear and collects the information together.*
>
> *You can do experiments that show these properties using trolleys and ticker timers to measure their velocity.*

CHECKPOINT

A space station of mass 30,000 kg is hit by a shuttle of mass 5,000 kg that is moving 0.1 m/s faster than the space station. If they stay joined together, by how much will the speed of the station change?

✦ **Force, Mass, Momentum, Newton's laws**

CONSTELLATION

This is a pattern of stars that we see in the sky. A lot of constellations have names, and they are useful for finding particular stars. The stars in a constellation are usually not really a group at all and are at huge distances from each other. They just happen to be in a similar direction from our Solar System. One example in the northern hemisphere is Ursa Major (the Plough or the Great Bear).

✦ **Galaxy, Star**

CONVECTION

This is the main process by which heat is transferred through *liquids* and *gases* (fluids). When part of the fluid is heated it expands and becomes less dense. It will therefore float upwards within the rest of the fluid, taking heat energy with it. Colder fluid takes its place and the process is repeated, which results in currents of rising warmer fluid and falling colder fluid. This can be seen in the apparatus in the first figure. The potassium permanganate crystal dissolves in the water to give a deep purple solution, and this is carried around by the convection current, making it visible.

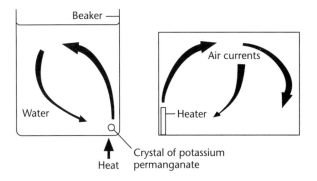

Beaker

Air currents

Water

Heater

Heat

Crystal of potassium permanganate

The reverse process is sometimes noticed near windows, where the fluid cools and becomes more dense so that it sinks. You may feel the cold air falling over the window sill on a cold day.

Room and caravan heaters are often convectors because it is a good way to spread the heat around a room without an exposed flame (see figure). A very large room will have the process speeded up by fans which blow some of the warm air around the room. This is called 'forced convection'. The hot water system in your house may work by hot water rising to the storage tank and being replaced by cooler water from the bottom of the tank.

> *Remember:*
>
> *(1) A gas is a poor conductor and can be used as an insulator if you trap it in thin layers or small pockets to stop convection (see **insulator**).*
>
> *(2) If you are asked how convection works it is NOT because 'heat rises'.*

✦ **Conduction of heat, Infrared radiation, Vacuum flask**

COULOMB

One coulomb is the charge transferred when a current of 1 amp flows for 1 second.

✦ **Charge**

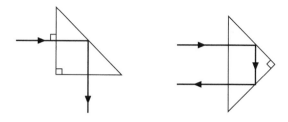

CRITICAL ANGLE

The critical angle is the angle of incidence when a ray of light leaving a denser material has an angle of **refraction** of 90°.

When a ray of light leaves a denser material it is *refracted* away from the normal so that the *angle of refraction* is larger than the *angle of incidence* (see the diagram).

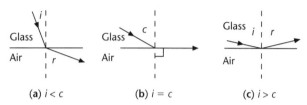

(a) $i < c$ **(b)** $i = c$ **(c)** $i > c$

If the angle of incidence is measured until the angle of refraction becomes 90°, the angle of incidence is the *critical angle* (see (b) in the diagram). If the angle of incidence is increased still more, something different must happen, as the angle of refraction cannot get any bigger. The ray is therefore reflected inside the surface and obeys the laws of **reflection** (see (c) in the diagram). This is called *total internal reflection* and must always happen when the angle of incidence is greater than the critical angle. Remember that this can only happen for a ray that is *leaving* a denser material. When a ray is *entering* a denser material it is always refracted.

To understand most of the uses of this effect you need to know the critical angle for materials such as glass and perspex. This can be measured using a semicircular block of the material, as in the following figure. The light must pass through M, the centre of the straight edge, so that it passes through the curved edge without changing direction.

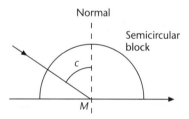

The critical angle for glass is 42°, so a ray of light hitting the inside surface of a piece of glass in air at an angle of incidence greater than this will be totally internally reflected. There are a number of uses for internal reflection, which are illustrated in the next figure. Make sure you can see why the ray is being reflected in each of the diagrams. Imagine a normal at each reflection and see that the angle of incidence is greater than 42°.

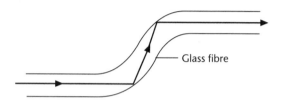

(a) Turning a ray through 90°   **(b) Turning a ray through 180°**

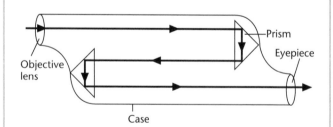

(c) Total internal reflection inside an optical fibre

Light is reflected through 90° in figure (a). Two of these prisms can be used to make a periscope. Light is reflected through 180° (as in figure (b)), inside the 'cat's eyes' on roads, through two prisms like those in figure (b) save some space. The back of the red plastic in bicycle reflectors and car rear lights also have lots of these small prisms.

Bundles of optical fibres are used to carry digital information in the telecommunications industry. Others can carry pictures from inside patients in hospitals. (Some of the fibres carry a light beam, while others bring back the reflected light. The fibres are so fine that their tiny dots of light make up the picture.)

Two glass prisms are used in each half of a pair of binoculars to fold up the light beam and make the telescope smaller (see the following figure). In real binoculars, one of the prisms is also rotated by 90°, which turns the final image so that it is the 'correct way up'.

+ Refraction

CURRENT

An electric current is a flow of **electrons** around a circuit.

We measure the electric current as the rate of flow of **charge** around the circuit. The unit for this is the **ampere** (A).

You should remember that there must be a force, the **electromagnetic force** (e.m.f.), to drive the current *all* the way around the circuit. The battery or power supply that provides this driving force is the source of the electrical energy. The circuit must be complete. If it is not, no current flows anywhere in the circuit.

The tiny electrons that carry the electric charge around the circuit cannot be created or destroyed, and the total current that enters any point on a circuit will be the same as the total current that leaves it. This is sometimes called Kirchhoff's first law and should help you to understand the circuits in the figure. An electric current is usually given the symbol I.

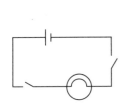

(a) Both switches must be on for the bulb to light

(b) The current is the same all the way round the circuit

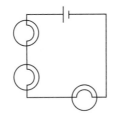

$$I_1 = I_2 + I_3$$
$$I_1 = I_4$$

(c) Kirchhoff's first law

Since the electrons are negatively charged, an electric current consists of electrons drifting slowly along the wire in very large numbers from the negative side of the supply around to the positive. This was not known when most of the rules about how electric currents behave were discovered, and the rules often refer to the *conventional current*, which is imagined to flow from positive around to negative. This is the direction of current referred to in this book.

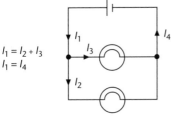

Correct

Incorrect

If you wish to measure an electric current you will need to use an ammeter (or a milliammeter for smaller currents). The meter must be placed in **series** with the part of the circuit that you are checking, so that the current flows *through* it, and have current flowing through it from its + terminal to the – terminal (see figure). This will mean that the

ammeter must have a very small **resistance** or it will change the current in the circuit when you connect it!

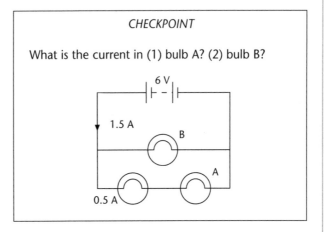

CHECKPOINT

What is the current in (1) bulb A? (2) bulb B?

6 V

1.5 A

B

A

0.5 A

⊸⊦ *Alternating current, Direct current, Ohm's law, Potential difference*

CURVED MIRRORS

There are two types of curved mirror. The type that curves towards the light is called *convex* and the type that curves away from the light is *concave*.

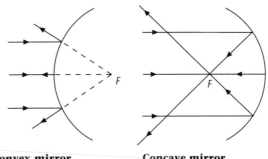

Convex mirror   **Concave mirror**

Parallel rays of light will be diverged by a convex mirror as though they had come from its *focus*, *F*. Parallel rays of light will be converged by a concave mirror so that they pass through the focus *F*. In both cases, the distance from the focus to the centre (pole) of the mirror is called the *focal length*. Remember that light can follow the same path in both directions – light from the focus of a concave mirror will emerge as a parallel beam.

The image produced by a convex mirror is always the correct way up (erect), smaller than the object, and **virtual**. It is mainly used in cases where you need a wide angle of view – some driving mirrors, mirrors in buses, large mirrors to watch for shoplifters in supermarkets.

The type of image produced by a concave mirror depends on the distance of the object from the mirror, but most uses depend on its ability to collect light together or to produce a parallel beam. Microwave dishes and radio telescopes use a large concave reflector to focus the parallel beam from a distant source on to an aerial at the focus. The large area of the mirror helps to collect enough energy from weak sources and bring it together onto the aerial. Polished metal reflectors behind the bulbs of torches and car headlights help to throw more energy forward in the correct direction. A more parallel 'long distance' beam is produced if the filament of the bulb is close to the focus of the mirror. If an object is closer to a mirror than its focus the image produced will be erect, magnified and virtual. You use this sort of image when you use a make-up or shaving mirror.

> *Remember: It is not only light that can be reflected by a mirror of this type. We have already mentioned microwaves and radio waves. Infrared (heat) radiation is also reflected forward by the metal mirror behind the element of an electric fire.*

A large, accurately made mirror such as that used in a telescope is not quite as simple in shape as those in the diagram. To get an accurate focus the outer edges are more 'flattened'. A mirror of this shape is called a *parabolic* reflector.

Real image, Reflection, Ripple tank, Virtual image

DECIBEL

This is the unit that is used to measure the level of sound or noise.

You will only just be able to hear a sound of 0 dB, but sounds of 120 dB or more can permanently damage your ears.

✧ *Noise, Sound*

DENSITY

Density is the **mass** of one unit of volume of a material.

$$\text{density} = \frac{\text{mass}}{\text{volume}}$$

$$D = \frac{m}{v}$$

We usually deal with mass in m³, so the units are kg/m³.

For some materials these units may not be very convenient. **Gases**, for example, may have their density given in g/l, but the SI units are kg/m³ and should be used whenever possible. The table below may help you to make sure that your answers for density are realistic.

> Remember: Materials that are less dense than water will float in it.
>
> Materials that are more dense than water will sink in it.

In finding density, the mass of a **solid** or **liquid** can be obtained by using a balance, subtracting the mass of the empty, dry container if necessary.

● The volume of a *regular solid* can be found by measurement and calculation, especially if the shape is simple.

● The volume of a *liquid* can usually be found by using a measuring cylinder (and do not forget to find the mass of the cylinder first so that you do not need a second, empty dry container!).

● The volume of a *powder*, especially a coarse one such as sand, poses problems because of the air trapped in it. Find the mass first (it will be very difficult when wet!), then add it slowly to a measured volume of water in a measuring cylinder. The increase in volume is the value that you want.

● The volume of an *irregular object*, such as a stone, can be measured by the displacement of water. For a small object, drop it into a measured volume of water in a measuring cylinder and note the increase in volume. A larger object will need a eureka (displacement) can (see figure). The can is first filled above the level of the spout and placed on a level surface for the excess water to drain away. An empty measuring cylinder is placed under the spout and the object is slowly and gently lowered in. The volume of the object is the same as the volume of the water displaced into the cylinder.

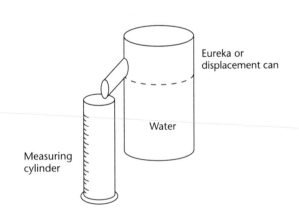

Eureka or displacement can

Water

Measuring cylinder

Material	Density in kg/m³	Material	Density in kg/m³
Water	1,000	brick	2,300
Concrete	2,400	wood (oak)	600
Iron	7,870	mild steel	7,860
Aluminium	2,710	lead	11,340
Gold	19,300	ice	920
Nylon	1,150	polythene	950
Olive oil	920	air	1.29
Carbon dioxide	1.98	hydrogen	0.090
Oxygen	1.43	methane	0.717

CHECKPOINT

A large steel girder has a volume of 0.25 m³. If the density of the steel is 8,000 kg/m³, what is the mass of the steel?

If the strength of the Earth's gravitational field is 10 N/kg, what does the steel weigh?

-+- **Mass**

DIFFRACTION

Diffraction occurs when a **wave** is made to spread out by passing through a narrow gap.

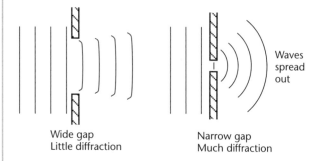

Wide gap
Little diffraction

Narrow gap
Much diffraction

The narrower the gap compared with the wavelength of the wave, the more it is diffracted; for example, red light would be diffracted more than green light at the same narrow slit.

-+- **Ripple tank, Wave**

DIFFUSION

This is the process where **gases** or **liquids** mix because of the free, random movement of their particles.

Solids will not diffuse through each other, because their particles are held in position in a crystal lattice.

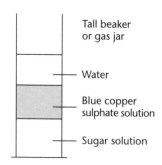

Tall beaker or gas jar

Water

Blue copper sulphate solution

Sugar solution

Liquids diffuse, but are fairly slow to do so because there are no spaces between the molecules to allow rapid movement. In the experiment shown in the figure it will probably take about a week before the colour is a uniform blue. If you wish to do this experiment you should put the water into the jar first and then put the other liquids under it, using a plastic syringe and a piece of rubber tubing.

Gases will diffuse rapidly because of the speed of their particles and the spaces between them. Even in a room with no draughts, a bad smell will quickly be noticed at the other side of the room. The experiment shown in the second figure has a piece of cotton wool soaked in ammonia solution at one end. As the ammonia gas diffuses along, it dissolves in the water on the damp litmus and turns it from red to blue. The whole process takes only a few minutes.

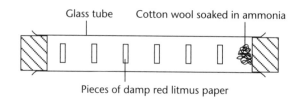

Glass tube   Cotton wool soaked in ammonia

Pieces of damp red litmus paper

-+- **Kinetic theory**

DIGITAL

A digital reading is one that increases or decreases in distinct small 'steps'. Only some values can be read, and readings between these values cannot be obtained. The system is quite acceptable provided that the 'steps' are small in comparison with the size of the reading. A digital watch has a display on which the reading changes in small steps, usually each second. Most computers process digital inputs and give digital outputs.

-+- **Analogue**

DIODE

A diode is a **semiconductor** device used in electrical and electronic circuits that allows current to pass in only one direction.

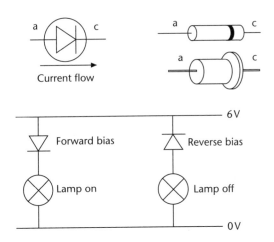

The circuit symbol and the two types of diode are shown in the figure above. The larger diode can carry a larger **current** than the smaller one without being damaged. The diode can carry a current in the direction shown by the arrow in its symbol. It is made of a p-type semiconductor and an n-type semiconductor, which meet at a junction. Current can get through the junction only when the p-type material is at a more positive voltage than the n-type. The wire to the p-type material is the **anode** ('a' in the diagram) and the wire to the n-type material is the **cathode** ('c' in the diagram). The device will have its symbol marked on it to show which wire is which or have a coloured band painted round it at the cathode end.

A diode that is connected so that it is conducting current is 'forward-biased'. If it is connected the other way round it is 'reverse-biased'. A diode can be investigated in the circuit shown in the figure below. Changing the variable resistor changes the current through the diode and the p.d. across it, so that you can get a set of readings from the meters. Plotting these on a graph gives the 'diode characteristic'.

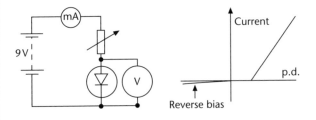

The reverse bias part of the graph can be checked by reversing the battery connections. A perfect diode would have no resistance in forward bias and infinite resistance in reverse bias. Real diodes are not as good as this. A silicon diode will need a forward bias of about 0.6 V before it conducts, and it does have a small resistance. It will also conduct a very tiny current when reverse biased, but this is quite acceptable in normal use. You should remember that the graph shows that the diode does not obey Ohm's law – i.e. it does not keep to a straight line.

-+- *LED, Rectifier, Relay, Semiconductor*

DIRECT CURRENT (D.C.)

Direct current always flows in the same direction around a circuit. The most common way of supplying this is from a battery. The voltage driving the current is usually constant, but many power supplies are not really able to provide this, especially when the current is large, and the current is then direct but not 'smooth' (see **Rectifier**).

-+- *Alternating current, Current*

DISPERSION

-+- *Spectrum*

DISPLACEMENT

This is the distance moved in a particular direction. Note that this is a **vector** quantity, and the direction should be either very obvious (movement along a straight track, for example) or clearly stated, so that the displacement is not mixed up with a simple distance, which is a **scalar**.

-+- *Scalar, Vector, Velocity*

DISTANCE/TIME GRAPHS

These graphs give you a clearer picture of the way in which an object is moving than you could see from a table of data. Most such graphs are really displacement/time graphs, because the movement takes place in a straight line, such as along a straight track.

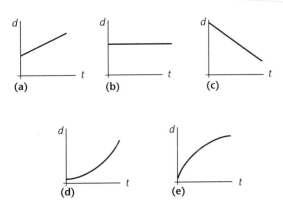

Look carefully at each part of the figure and make sure that you understand what the shape of each graph tells you. All of them show the distance moved by a railway truck along a railway line. In (a), it is moving at uniform **velocity**; in (b), it has stopped at some point along the track; in (c), it is coming back along the track at uniform velocity; (d) shows the truck accelerating; and (e) shows it decelerating.

You can also find the velocity of the object from a straight-line graph. Read the change in distance from the distance axis and the time taken from the time axis. Put the numbers into

$$\text{velocity} = \frac{\text{change in displacement}}{\text{time taken}}$$

This is the same as finding the *gradient* of the graph.

Worked example

Find the velocity of the truck in the figures below. You should have found that

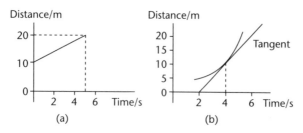

(a) (b)

$$\text{velocity} = \frac{\text{change in displacement}}{\text{time taken}} = \frac{20 - 10}{5}$$

$$= 2 \, \text{m/s}$$

$$\text{velocity at 4s} = \text{gradient of tangent} = \frac{20}{4}$$

$$= 5 \, \text{m/s}$$

CHECKPOINT

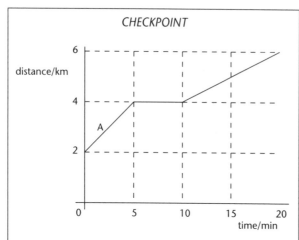

A cyclist's journey

From the graph find:
(1) The speed of the cyclist in part A of the journey.
(2) How long the cyclist was stopped.
(3) The average speed for the whole journey.

> *Remember: A line that slopes down from top left to bottom right shows a negative speed – the object is moving back towards its starting point.*

⫶⫶ Displacement, Speed, Velocity, Velocity/time graphs

DOMESTIC ELECTRICITY

The electricity supply for use in the home is supplied via the **National Grid System**. In order to save energy by transmitting the **power** at high voltages the system uses **transformers**, so the supply is always a.c. The domestic supply is usually 240 with a frequency of 50 Hz. The supply enters the house through a main **fuse**, which belongs to the Electricity company and is sealed by it. Since this cable is permanently live and has little current limitation the fuse is *not* repairable by the consumer. The supply then passes through a meter, which shows the energy used so that the bill can be worked out each quarter (see **Payment for electrical energy**). The cable is then connected to the main distribution board. A suitable **earth** will also be connected to the board. The board may carry miniature circuit breakers instead of fuses or a residual current circuit breaker (RCCB) or earth leakage circuit breaker (ELCB – see **Earth**). There will be a lighting circuit for each floor of the house, and each light and its switch will be connected in parallel in this circuit. A lighting circuit will carry a 5 A fuse.

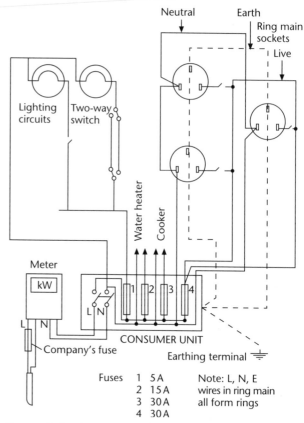

Domestic 'mains' installation

There will also be a 'ring main' for each floor of the house. Each of the sockets, with a 13 A maximum rating, is connected in parallel across the ring main and both ends of the 'ring' are connected to the supply. This means that every socket has two cables connecting it to the supply, and the chance of overheating the cable is reduced. Each ring main will be protected by a 30 A fuse. In addition to these there will be a supply to the cooker, which takes so much energy that it has its own circuit and a 30 A fuse, and, if required, an immersion heater circuit with a 15 A fuse.

-⁜- *Alternating current, Earth, Fuses, National Grid System*

DYNAMO

A dynamo is a device that charges **kinetic energy** into electrical energy. The kinetic energy is used to rotate a coil in a magnetic field (or to rotate a magnet in a coil). An e.m.f. is then generated in accordance with **Faraday's law**. If the output is a.c. rather than d.c., the dynamo is called an **alternator**.

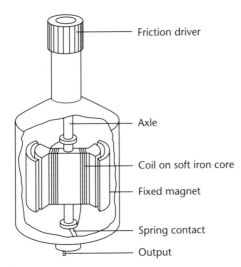

A bicycle dynamo

One form of alternator is the *bicycle dynamo*. As the coil is rotated on its axle the field through it changes continuously, and an e.m.f. is induced. This electrical output stops when the bicycle stops – no energy input – and increases as the bicycle is ridden faster – greater **power** input. If you connect the output to an oscilloscope it is clearly shown to be a.c. This happens because the movement of the coil in relation to the fixed permanent magnet is reversed every half a revolution, first approaching an N pole and then a S.

Sometimes the coil and magnet poles are drawn as in the figure below, where it is easier to apply **Fleming's right-hand rule** and find the direction of

the current. If you apply Fleming's right-hand rule to side A the direction of the current is as shown in the diagram. Half a turn later, side A will be moving down in the field instead of up and the direction of the current in the outside circuit, R, is reversed. The output is therefore a.c. The maximum output will be obtained when the field, movement and current are all at 90°, and the output will be zero when the wire moves parallel to the field (a quarter turn after the position shown in the diagram). This can be seen on the diagram of the voltage output.

If a d.c. output is needed, a 'split-ring' commutator can be used instead of the complete rings used for a.c. The splits pass the brushes at the same instant that the current reverses direction and the current in the outside circuit always flows the same way – although it is *not* a constant voltage like the output of a battery.

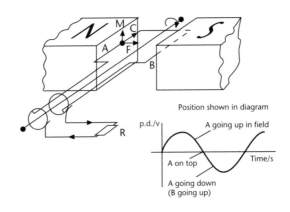

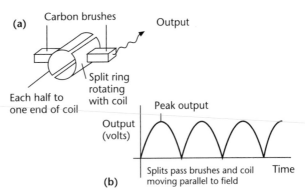

A split-ring commutator

-⁜- *Electromagnetic induction, Faraday's law, Fleming's right-hand rule, Lenz's law*

EARTH (ELECTRICAL)

All electric appliances with metal parts that can be touched should have those metal parts connected to earth to prevent electric shocks.

The earth is at 0 V, so any voltage above this will try to drive current down to earth in any way that it can. If a live wire in a mains appliance accidentally touches an outside metal part and you touch the part the current will go through you. Since your muscles, including your heart, are controlled by tiny electric signals, the electric shock can be fatal.

If the metal part is connected to earth by an earth wire then the current will go through the cable instead of you. The current is usually big enough to 'blow' the fuse and cut off the current.

The earth wire is usually coloured green and yellow. It is connected to the largest pin on 13 A plugs so that it is connected first.

If a good earth is not possible, then an ELCB (earth leakage circuit breaker) can be used. It compares the current in the live and neutral wires near to the supply. If there is more than a tiny difference between the two, then it switches off the supply, because some current must be going to earth. This sort of device would be used in a caravan, where no earth is available.

Some appliances are specially made so that it is impossible for live wires to connect to any metal part that you could touch. These are called double-insulated and do not need an earth wire. They will have the following symbol marked on them:

-+- **Domestic electricity, Fuse, Potential difference**

EARTH (PLANET)

-+- **Solar system**

EARTHQUAKES

Earthquakes happen because the boundaries of the plates that form the surface of the Earth move relative to each other. Huge frictional forces stop this happening for a while until enough energy is built up for a sudden movement.

The place in the Earth's crust where the earthquake takes place is the *focus*, and the place on the Earth's surface directly above the focus is the *epicentre*.

Waves are generated by the movement and travel outwards from the focus. They can be detected by a *seismometer*. There are three types of wave:

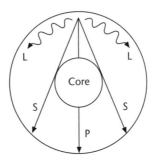

- ● *P waves* These pressure waves are compression waves that travel rather like sound waves. They can travel through solids and liquids and will reach the opposite side of the Earth.

- ● *S waves* are transverse (shear) waves which cannot travel through a liquid. They are reflected at the liquid outer core of the Earth which therefore leaves a shadow. This allows scientists to work out the size of the core.

- ● *L waves* are long waves that travel around the crust. These arrive last but do the most damage.

-+- **Wave**

EFFICIENCY

$$\% \text{ efficiency} = \frac{\text{work output}}{\text{work input}} \times 100\%$$

Machines convert **energy** from one form to another. Most devices that do this are not very efficient because a lot of the energy that goes in does not end up where you want it to be! In many cases some of the energy goes into heat which is 'lost' into the air. This can happen with devices as different as a car engine and the heat sink in an amplifier. Remember that the energy is not really lost or destroyed – it has ended up in a form that you do not find useful.

The equation could also have been written as

$$\% \text{ efficiency} = \frac{\text{energy output}}{\text{energy input}} \times 100\%$$

or

$$\% \text{ efficiency} = \frac{\text{power output}}{\text{power input}} \times 100\%$$

The efficiency is sometimes left as a fraction, but it is more useful to change it into a percentage so that different machines can be compared more easily. It is a good way to compare devices that do the same job, but there may be other factors that you may think are more important, such as the fuel used, pollution caused, size of the machine, etc. If an electric car is more efficient than a petrol car, is it always the best car to use?

Systems based on burning fuels, such as car engines, are not very efficient – about 30 per cent – but some electrical systems such as transformers can be quite close to 100 per cent.

Worked example

A crane motor is 35 per cent efficient. If the energy used in lifting a load is 70,000 J, how much energy is supplied to the motor?

$$\% \text{ efficiency} = \frac{\text{energy output}}{\text{energy input}} \times 100\%$$

$$35 = \frac{70,000}{\text{energy input}} \times 100$$

$$\text{energy input} = \frac{70,000}{35} \times 100$$

$$= 200,000 \text{ J}$$

CHECKPOINT

A crane uses 8,000 J of electrical energy to lift a load onto a new building. If the load gains 5,000 J of potential energy, how efficient is the crane?

-⊹- *Energy, Power, Machine, Work*

EFFORT

This is the name used for the force being put into a machine.

-⊹- *Lever, Load, Machine*

ELASTIC LIMIT

-⊹- *Hooke's law*

ELECTRIC MOTOR EFFECT

A *force* will be produced on a conductor carrying a *current* in a *magnetic field*.

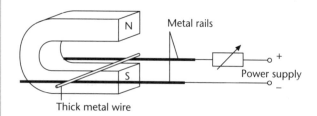

Thick metal wire

The apparatus shown in the figure above shows how to demonstrate this force. With the circuit wired as shown, the thick wire will roll along the metal rails towards the right. If the direction of the current

is reversed, the wire rolls the other way. The force can be seen to increase if a stronger magnet is used or the current is made larger (the wire then rolls faster).

Notice that the apparatus has been set up to have the motion, field and current all at 90° to produce the largest effect. The direction of the force can be found from *Fleming's left-hand rule*. The force is produced by the interaction of the two magnetic fields, one from the magnet and the circular one produced by the current in the wire (see figure). The result is sometimes called a 'catapult field'.

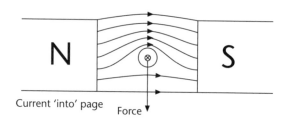

Current 'into' page   Force ▾

A simple d.c. motor can be made from a coil that can rotate on an axle between the poles of a permanent magnet, as shown in the figure below. Trace the current around the circuit and apply Fleming's left-hand rule to each side of the coil, X and Y. X will be moved up and Y will be moved down. They would then stop after a quarter of a turn, when they had gone as far up or down as possible. At this time, X must stop going up and start to come down if the rotation is to continue. Y must also reverse its direction. The simplest way to do this is to reverse the direction of the current by using a *split-ring commutator*. Each end of the coil is connected to one half of a split copper ring that rotates along with the coil. The current gets into the coil through the brushes that rub on each side of the split ring. Each time that the current must change direction the splits pass the brushes, so the connections from coil to power supply are reversed. The coil can then continue its rotation.

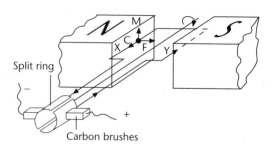

Split ring

Carbon brushes

The 'turning effect' from this simple motor can be increased by using a magnet of greater field strength, a coil with more turns or a greater current.

The same principle can be used to measure small electric currents in a *galvanometer*. In this case you do not need the split-ring commutator, because the coil will turn through less than a complete revolution. However, it will turn against a spring, so that a greater current can produce a greater turning force and rotate the coil through a greater angle. A pointer attached to the coil will then be moved along a scale, which can be calibrated in mA.

✦ *Fleming's left-hand rule, Loudspeaker*

ELECTROMAGNET

A simple electromagnet has a coil of wire wound around a core. When a current is sent through the coil, the electromagnet behaves like a bar magnet. The core will be made from a magnetic material which is easy to magnetize.

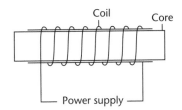

An electromagnet

The magnet can be turned off by switching off the current and can be made stronger by putting more turns on the coil or using a bigger current.

The magnetic core greatly increases the effect and the magnet is much weaker if the coil is wound on an empty tube.

All of these effects can be investigated by counting the number of paper clips that can be picked up by an electromagnet made from a coil of wire wound around a nail and powered by batteries.

Which end of the magnet is N and which is S depend on the direction of the current – see ***Magnetic poles and magnetic forces***.

✦ *Magnetic fields, Magnetic materials*

ELECTROMAGNETIC INDUCTION

An ***electromotive force*** (voltage) is produced when the magnetic field linked with a circuit is changed.

When the wire in the figure below is moved down through the field, the magnetic field through that loop of the circuit changes – we say that there is a change of 'flux linkage'. This induces an electromotive force (e.m.f.), which drives a current in the direction shown, and this can be measured on the **galvanometer**. The direction of a current induced in this way can always be found from **Fleming's right-hand rule**.

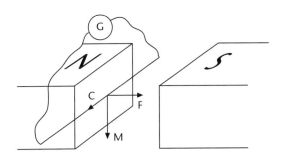

The e.m.f. (and therefore the current) can be increased by moving the wire faster or by using a stronger magnet. The e.m.f. can also be increased by winding the wire into a coil – the flux linkage is then increased by a factor equal to the *number of turns*. The direction of the current is reversed if the direction of motion is reversed or if the poles of the magnet are the other way round. In a similar way to the electric motor effect, the strongest effect will be produced when the field, motion and current are all at right angles.

A similar effect can be obtained by pushing the pole of a magnet into or out of a coil (see figure). As the magnet moves there is a change in flux linkage; the *faster* the flux linkage changes the *greater* the e.m.f. produced.

If you try this, you will need a sensitive galvanometer because the e.m.f. produced will probably be quite small. If you carefully change one factor at a time, you will notice that:

- When the magnet is not moving, no e.m.f. is produced.

- The faster the magnet moves, the bigger the e.m.f.

- A stronger magnet produces a bigger e.m.f.

- A coil with more turns produces a bigger e.m.f.

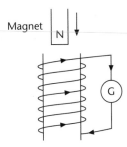

These results and those from the investigation about moving a wire/coil in a magnetic field are summarized in **Faraday's law**.

If you check the direction of the current produced when the magnet is moved into and out of the coil you will also find that:

- Reversing the direction of movement reverses the induced e.m.f.

- Using the other pole of the magnet also reverses the e.m.f.

Checking the magnetic pole produced by the current in the coil (see ***Magnetic poles and magnetic*** *forces*) shows that the pole always *opposes* the movement of the magnet – it repels the pole of an approaching magnet and attracts the pole of a magnet moving away. This is an example of **Lenz's law,** which can usually help you to find the direction of an induced current.

The playback head on a tape recorder uses this induction effect to retrieve information from a recorded tape. The 'tape head' is a coil and, as the magnetic pattern on the tape moves past it, a current is induced in it. This current is then increased in an amplifier and used to drive a **loudspeaker** (see also **Magnetic poles and magnetic forces**).

Another very important use is in **dynamos** and generators. Many modern dynamos produce a.c. and are called **alternators. Transformers** also use the same effect.

-╬- *Dynamo, Faraday's law, Fleming's right-hand rule, Lenz's law, Loudspeaker, Magnetic poles and magnetic forces, Transformer*

ELECTROMAGNETIC SPECTRUM

The most familiar of the waves in the magnetic spectrum is *light,* but they all have many similar properties and a lot of common uses. They all have the properties of **waves**. All these waves consist of a vibrating electric and magnetic field and do *not* need a material to travel in, so they can reach us through space, as heat and light do when they come from the Sun. All of the waves move at the same very high speed in a vacuum, about 300,000,000 m/s – written as 3×10^8 m/s. They slow down in other materials. Going *down* the list, the **wavelength** gets shorter, the wave carries more energy, and it is usually more penetrating. There is no sudden change from one group of waves to the next and the names refer to approximate areas of the spectrum. The figures for wavelength are only an approximate guide.

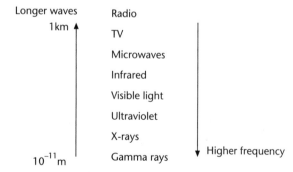

The electromagnetic spectrum

Radio waves

These will have a wave length of about 1 km down to a few metres and are transmitted and received by electronic equipment connected to an aerial – although it is true that a spark sends out radio waves and creates a common form of 'interference'.

Microwaves

These have a wavelength of about 1 cm and are used to transmit telephone calls between towers carrying dish aerials as well as to heat food in microwave ovens. A microwave oven shows that the waves carry energy and that they are quite penetrating, as the waves are absorbed inside the food and the energy turns to heat. Metal objects are not allowed in the oven because they can reflect the waves, damaging the oven.

Infrared waves

These have a wavelength from about 1 mm down to the red end of the visible spectrum. They are given out by hot objects (including you). The radiated heat from an electric fire consists of infrared waves, which change into heat energy when they are absorbed. Hotter objects can radiate shorter wavelengths. You can detect infrared using photo-diodes and send them from special **LEDs** (used in TV remote-control handsets). The infrared that your body emits will trigger security sensors and show up on infrared gun sights.

Visible light waves

These extend from the red to the violet at a wavelength of about 5×10^{-7} m (see **Spectrum**). Visible light is emitted from the electrons of atoms that have a lot of energy in short bursts, called photons. Detectors include the eye, **LDR** and photo-electric cells.

Ultraviolet waves

These are shorter in wavelength than the visible spectrum, at about 10^{-8} m. They carry more energy and cause skin cancer as well as sunburn. They are given out by certain fluorescent light tubes that contain mercury and make bad eggs glow with a slightly different colour from fresh eggs. They will also make some detergents fluoresce (glow) a fairly bright blue.

X-rays

These come from the electrons of atoms that have been hit by high-speed electrons and have a wavelength of about 10^{-10} m. They are quite penetrating and affect photographic film, which gives them their medical use. Large or very frequent doses can be dangerous and lead to skin or internal cancers.

Gamma rays

These are similar to X-rays except that they come

from the nuclei of radioactive atoms. They are usually shorter in wavelength and correspondingly more dangerous, since they carry more energy and are more penetrating (see **Radioactive decay**). Cosmic radiation reaches us from space and includes a variety of wavelengths, some of which are in this dangerous and penetrating range.

-+- **Infrared radiation, Gamma radiation, Spectrum, Waves**

ELECTROMOTIVE FORCE (E.M.F.)

Electromotive force is a measure of the driving force that drives the **electrons** all the way around an electric circuit (including the supply).

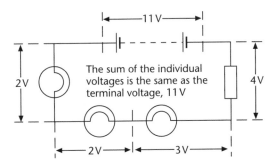

The sum of the individual voltages is the same as the terminal voltage, 11 V

The e.m.f. is usually given the symbol E and is measured in **volts**. The e.m.f. in the circuit will be divided up, so each component has the necessary voltage to drive the **current** through it. If you *add together* all the voltages round the circuit, you will find that they add up to the *terminal voltage* of the supply (see figure). The rest of the e.m.f. is used to drive the current through the supply itself. The individual voltage across each part of the circuit is called the **potential difference**. So a more accurate definition is that the e.m.f. is the total of all the potential differences in a complete circuit.

CHECKPOINT

If a set of Christmas tree lights has twenty 12 V bulbs in series, what voltage should the supply be?

-+- **Current, Potential difference**

ELECTRON

An electron is one of the three particles from which **atoms** are made. It has a negative **charge** and almost no **mass** (about 1/1840 of the mass of a **proton**). It is found in energy levels *outside* the nucleus of the atom, which are sometimes pictures as *orbits*.

-+- **Atom, Neutron, Proton**

ELECTROSTATICS

-+- **Static electricity**

ELEMENT

Elements are substances that contain only *one* type of atom. Their particles may be single **atoms** or **molecules**. For example, hydrogen is an element; hydrogen gas contains hydrogen molecules. Copper is also an element; copper contains copper atoms.

Elements cannot be broken into simpler chemical substances.

ENDOTHERMIC REACTION

Energy is always involved in a chemical reaction; the substances reacting sometimes *take in* energy and sometimes *give out* energy. Endothermic reactions are those in which energy in the form of heat is transferred from the surroundings to the substances; in other words heat is taken in.

-+- **Exothermic reaction**

ENERGY

Work is done when energy is transferred from one system to another. Energy may be contained in different ways in different systems. For example, chemical energy in a battery is transferred to electrical energy when a current flows, and this may be transferred to light and heat energy in a bulb. When this happens energy is neither created nor destroyed, so the total energy after the conversion is the same as the total energy before. This is the principle of *conservation of energy*.

Since we can turn the different forms of energy into work we use the same unit for energy, the **joule**, as we do for **work**. Quantities of energy are calculated by finding how much work has been done or could be done.

work done = energy transformed
power = rate of conversion of energy

Energy is transferred from one place to another by a **wave**, and we often use the waves of the **electromagnetic spectrum** to do so. For example, the walls of a room are warmed by the heat produced when the infrared waves from an electric fire are absorbed.

All the following forms and sources of energy have an entry in this book: **chemical, nuclear, kinetic, potential, internal, solar, tidal, wave, biomass, geothermal.**

Machines convert energy from one form to another, and we can often trace the energy through a series of *different forms* within a complete process. For example, in a **hydroelectric** power station, the water has **gravitational potential energy**. As it falls, the energy is converted into **kinetic energy**, which is used to drive the turbines. The rotating shaft from the turbine now carries the kinetic energy into the alternators (see **Dynamo**), where it is converted into electrical energy. None of these machines will be perfect, and the final quantity of electrical energy will be *less than* the original potential energy. The *missing* energy has not been destroyed – it is simply not in the form that we would like it to be. In this case, most of it will have been 'lost' as heat to the surroundings. We measure how good machines are at converting energy by measuring their **efficiency**.

It is more efficient to *run* the generators in a power station at a constant rate. Unfortunately we do not *use* the electrical energy at a constant rate, and power stations sometimes produce more energy than we require and sometimes less. We can avoid this by *storing* the surplus energy in a pumped storage system. When too much electrical energy is being produced, it is used to pump water up into a higher reservoir, where it has a lot of gravitational potential energy. When we require more energy, the water is allowed to flow back down again and the energy is converted back into electrical energy by turbines.

Our bodies take in chemical energy as food (and store some of it as fat!). When the cells need energy they can burn some of this fuel by combining it with the oxygen that is transported by the blood stream. The waste products from the reaction are taken away by the blood stream and we breathe them out as carbon dioxide and water vapour, i.e. we use an **exothermic reaction**.

$$\text{glucose} + \text{oxygen} \rightarrow \genfrac{}{}{0pt}{}{\text{carbon}}{\text{dioxide}} + \text{water} + \text{energy}$$

⊹⊹ *Sources of energy*

EQUATIONS OF MOTION

These equations are true for objects that move with constant **acceleration**.

$$v = u + at \qquad s = ut + \tfrac{1}{2}at^2 \qquad v^2 = u^2 + 2as$$

where u = original velocity, v = final velocity, t = time taken, a = acceleration, s = displacement.

⊹⊹ *Acceleration*

EQUILIBRIUM

When a body is in equilibrium, all the **forces** on it are balanced, so that it is not accelerated in a straight line or a circle.

If it does not accelerate in a straight line then, for any parallel forces, the sum of the forces in one direction must be the same as the sum of the forces in the opposite direction. If the body does not

accelerate in a circle, the principle of **moments** must apply.

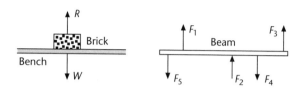

In the figure the brick must have an upward force, a 'reaction' equal in size to its weight, or it would accelerate downwards. The forces on the beam must balance, so that

$$F_1 + F_2 + F_3 = F_4 + F_5$$

or it will start to move.

We are also often interested in how stable bodies are. We can get an idea of the *stability* of a body if we look at the position of its **centre of mass** and its base. The weight of a body will behave as though it all acts downwards from the centre of mass and this can cause a moment that will make it topple over.

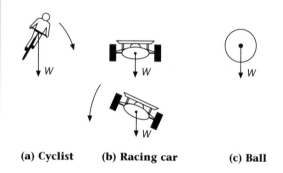

(a) Cyclist   **(b) Racing car**   **(c) Ball**

● (a) The stationary cyclist will be *unstable* because a slight movement to one side produces a moment that topples her over.

● (b) The racing car is much more *stable*. If it is tipped on to one wheel, the moment produced tips it back on to its wheels. It would have to go a very long way over for the weight to be outside its wheelbase and cause it to fall over.

● (c) The ball is in a state called *neutral equilibrium* because the weight always passes through it in the same way no matter which position it rolls into. In general terms, a wide base and a low centre of mass make things more stable.

> *Remember: an object can stay upright only if its centre of mass is above its base. Try standing your pencil on its point, end, side – what sort of stability have you got?*

⊹⊹ *Centre of mass, Force, Moment*

EVALUATION

This is the last part of writing up an investigation. Try to include answers for the following:

- Are there ways that you could have improved on your method, especially ways that would have made your measurements more accurate?
- If you did the investigation again, what improvements would you make?
- Where there any **rogue results**? If so, what did you do about them?
- Were all the results in a good pattern? Were they close to a line of best fit on the **graph**?
- Do your results allow you to be confident that your conclusions are correct?
- Did you test a wide range of values to check your prediction as far as you could?
- Did you do enough repeats to check the accuracy of your results?
- Are there other things that you could do as further work for this investigation?
- Have you looked up the theory and clearly sorted out the science behind the investigation?

⊹ **Graph, Rogue results**

EXOTHERMIC REACTION

Energy is always involved in a chemical reaction: the substances reacting sometimes take in energy and sometimes give out energy. Exothermic reactions are those in which energy in the form of heat is transferred to the surroundings; in other words, heat is given out.

⊹ **Endothermic reaction**

EXPANSION

Solids, liquids and gases will all expand when heated. **Solids** will expand least, but the forces when they do expand will be very large. **Gases** will expand most at constant pressure, but there are spaces between the particles and it is possible to increase the temperature at constant volume so that the **pressure** increases instead (see **Kinetic theory**).

> *Remember: You know that liquids expand when heated – you can see it happen in a thermometer.*

⊹ **Expansion of liquids, Expansion of solids, Kinetic theory**

EXPANSION OF LIQUIDS

We can easily show that the *volume* of a liquid increases with *temperature*, but it is a little more difficult to measure, because the container will also expand.

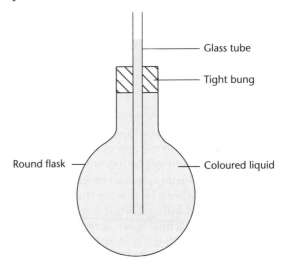

Liquids expand more than solids. The forces when they expand are quite large and can break sealed containers. This increase in volume is used in **thermometers**.

The expansion of water is not uniform, and this has important consequences. If you cool water from room temperature it will contract until 4 °C. Below this temperature it expands again until it reaches its freezing point. This means that below 4 °C the colder water floats to the top and water freezes from the surface downwards instead of from the bottom upwards (which is the case for most other liquids). If it were not for this odd behaviour, water-life would die in winter if ponds and lakes freeze.

⊹ **Expansion of solids, Thermometer**

EXPANSION OF SOLIDS

You can show the expansion of a metal with the simple apparatus shown in the figure. One end of the tube is clamped so that all movement of the tube occurs at the other end. This turns the pencil, which acts as a simple roller, and the movement is also magnified by the paper pointer. The pointer moves quite clearly when the steam is turned on. You can get an idea of how much the tube will expand from the first worked example below.

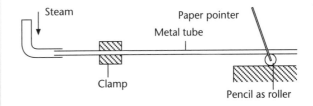

Many problems are caused by the large forces of linear expansion (and forces of the same size caused by contraction on cooling). Expansion gaps must be left between railway lines and at the ends of bridges or they may be damaged on very hot days. A glass dropped into hot water may break because it is not a very good conductor and its outside expands before the inside!

> *Remember: When a solid is heated, any holes in it get larger. (Think of the strip of metal around the edge of the hole, which must be getting longer.)*

Tightly screwed-on bottle caps can be loosened by running warm water on the cap so that it expands a little (the water should not be too hot or it may break the bottle!). Metal tyres can be fitted very tightly to railway wheels by making the tyre very slightly smaller and fitting it while hot. The only way to remove it is to machine it off. Rivets are often put in hot so that they pull the materials together when they cool. Other uses include the **bimetallic strip**.

Note that although we have measured only the linear expansion, a solid does expand in all directions, so its volume increases.

> *Remember: The forces when solids expand and contract are always large. If there is no gap to allow for this then damage will be caused.*

CHECKPOINT

How could you use expansion to fit a metal wheel very tightly on to an axle?

The expansion of **solids** is usually compared by a number called the *expansivity*. This is the fraction of its original length by which the solid expands for each degree increase in temperature.

$$\text{expansion} = \begin{array}{c}\text{linear}\\\text{expansivity}\end{array} \times \begin{array}{c}\text{original}\\\text{length}\end{array} \times \begin{array}{c}\text{rise in}\\\text{temperature}\end{array}$$

Worked examples

1. A steel rod is 2 m long at 0 °C. By how much will it have expanded at 100 °C? (Linear expansivity of steel = 0.000015/K)

   expansion = linear expansivity × original length × rise in temperature
   $$= 15 \times 10^{-6} \times 2 \times 100$$
   $$= 0.003 \text{ m}$$

   Remember that a temperature rise of 1 °C is the same as a temperature rise of 1 K.

2. A concrete bridge is 80 m long at −10 °C on a cold night in winter. How much will it have expanded on a hot summer day when the temperature is 30 °C? (Linear expansivity of concrete = 0.000012 K)

   expansion = linear expansivity × original length × rise in temperature
   $$= 0.000012 \times 80 \times 40$$
   $$= 0.038 \text{ m} = 38 \text{ mm}$$

The expansion is quite small but takes place with a *very* large force that will damage the bridge if there is no gap for it to expand into. Some linear expansivities for other materials are shown in the table below.

Material	Linear expansivity/K	Material	Linear expansivity/K
steel	0.000015	aluminium	0.000023
copper	0.000017	invar*	0.0000009
brick	0.000009	concrete	0.000012
glass	0.000008	PVC	0.000150

*invar is an alloy of iron and nickel.

Bimetallic strip, Expansion of liquids

FARAD

-+- *Capacitor*

FARADAY'S LAW

When the **magnetic field** linked with a circuit changes, an **electromotive force** is induced in the circuit. The size of the e.m.f. is directly proportional to the *rate of change* of the magnetic flux linked with the circuit.

-+- *Electromagnetic induction*

FISSION

When the **nucleus** of some large **atoms** is hit by a **neutron** the neutron is absorbed for a very short time and then the new nucleus splits into two parts, together with some more neutrons. This process of 'splitting the atom' is called nuclear fission.

The process is useful because the new nuclei that are produced require less energy to hold them together than the original large nucleus and the surplus energy is released as heat. If the masses of all the parts before and after the fission are checked very carefully, it is found that there is a small loss of mass, which has been converted into energy. This obeys Einstein's equation, which says

$$E = mc^2$$

Where E is the energy produced, m is the mass converted and c is the velocity of light. The value of c^2 is very large, so a small quantity of mass is converted into a lot of energy. Only certain nuclei will produce this effect. The best known ones are the nuclides ^{235}U and ^{239}Pu. Plutonium-239 does not occur naturally and has to be produced in a nuclear reactor from uranium-238. The fission of uranium-235 occurs as follows:

$$\begin{array}{c} 2 \text{ or } 3 \text{ neutrons} \\ + \\ ^{235}_{92}U + {}^{1}_{0}n \rightarrow {}^{236}_{92}U \rightarrow 2 \text{ nuclei} \\ + \\ \text{energy} \end{array}$$

It is then possible for some of the new neutrons to go on and hit other nuclei and cause them to split. If the process continues in this way it is called a *chain reaction*. If the mass of the fuel is very small, the neutrons leave it without hitting another nucleus and the reaction stops. If the mass is big enough, one neutron (on average) will hit another nucleus and cause it to split, so the chain reaction carries on. This mass, where the chain reaction is just maintained, is called the *critical* mass.

If the mass is bigger than the critical mass, more than one neutron hits another nucleus and the chain reaction starts to grow. Since the atoms are very close together, this will happen *very rapidly*, and in a very short time millions of atoms are splitting at the same time, releasing huge amounts of energy in a nuclear explosion. An 'atomic bomb' or 'nuclear weapon' will consist of two or more pieces of plutonium that are each below the critical mass. A conventional explosive forces these together so that the mass becomes much greater than critical and the chain reaction produces heat energy, extremely high temperatures, a pressure wave ('blast'), radiation, radioactive dust, etc. The use of such a 'simple' weapon would cause great problems for *both* sides in a conflict, as each reaction produces two 'new' nuclei. The nature of these cannot be predicted exactly, but many will be radioactive isotopes and there will be a great variety of **half lives**. Some will decay in a short time, producing dangerous ionising particles near to the scene of the explosion, but others will be decaying for years in dust which may be carried into the upper atmosphere, falling to the surface much later.

It is possible to control the reaction by controlling the number of neutrons that reach other parts of the fuel. This is done in a nuclear reactor by absorbing neutrons in *control rods* (see **Nuclear energy**).

-+- *Atom, Nuclear energy*

FLEMING'S LEFT-HAND RULE

This rule enables you to find the direction of the **force** that is produced when a conductor carries a **current** in a **magnetic field**. Put the first finger, second finger and thumb of your *left* hand at 90° to each other (see figure). Point the **f**irst finger in the direction of the field and the se**c**ond finger in the direction of the **c**urrent. The thu**m**b will then point in the direction of the **m**otion. Note that in this case you are looking at the *motor effect*, i.e. current is put in and motion is produced.

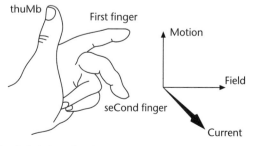

Fleming's left-hand rule

-+- *Electric motor effect*

FLEMING'S RIGHT-HAND RULE

This rule enables you to find the direction of the **current** that is produced when a conductor moves in a **magnetic field**. Put the first finger, second finger and thumb of your *right* hand at 90° to each other (see figure). Point the **f**irst finger in the direction of the **f**ield and the thu**m**b in the direction of the **m**otion. The se**c**ond finger will then point in the direction of the **c**urrent. Note that in this case you are looking at the *generator effect*, i.e. motion is put in hand and current is produced.

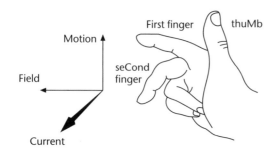

Fleming's right-hand rule

-**✦**- *Electromagnetic induction*

FOCUS

-**✦**- *Curved mirrors, Lens*

FORCE

Forces can pull, push, bend and twist things and are measured in newtons (N).

Forces can bend, stretch or 'deform' a material, and the amount that material is deformed may help us to find the size of the force, using **Hooke's law**. Stretching a spring or bending a ruler are good examples of this.

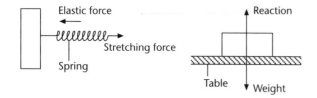

When a force is applied to a body that is not accelerating, there *must* be at least one other force to produce the 'balance' and make the total force zero. The body is then said to be in *equilibrium* (see figure). In the case of a spring, the force stretching it is balanced by the elastic force of the spring. In the case of a body standing on a table, the weight is balanced by an upwards force from the table, called a

reaction. A reaction will always be at 90° to the surface that produces it.

If a force is applied to an object that can move freely, it will accelerate in the direction of the force instead of being deformed. A larger force will produce a greater **acceleration**, but the effect will also depend on the mass of the object, which is a measure of its *inertia*. A more massive object will need a larger force to produce the same acceleration. More details on this and some worked examples can be found under **Newton's second law**.

Forces often cause things to rotate. When they do this in pairs they are called a *couple*. More details on turning forces can be found under **Moment**.

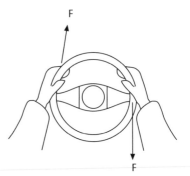

A couple acting on a steering wheel

-**✦**- *Centripetal force, Hooke's law, Newton's laws, Vector, Work*

FORCE METER (NEWTON METER)

-**✦**- *Hooke's law*

FOUR-STROKE ENGINE

The four-stroke engine is a heat engine in which the **chemical energy** of the fuel is turned into heat energy by burning it explosively with air, and then into **kinetic energy** as the crankshaft rotates. Its operation is best seen as four separate strokes, each taking half a rotation of the crankshaft. In each stroke, the piston will therefore move from top to bottom or bottom to top of its stroke.

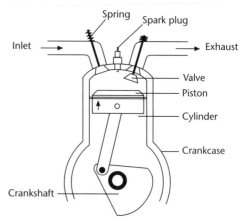

Four-stroke engine

Induction

During this stroke, the inlet valve is open and, as the piston moves down, a mixture of air and petrol vapour is drawn in through the valve from the carburettor.

Compression

Both valves are now closed and the piston moves up, compressing the petrol/air mixture into about a ninth of the original volume. Just before the top of the stroke a spark across the gap of the plug starts the mixture exploding.

Ignition

Both valves are still closed and the expansion of the exploding mixture forces the piston down. This is the only power stroke of the four strokes, and a flywheel on the crankshaft is used to keep the engine rotating through the other three strokes.

Exhaust

The exhaust valve is opened and the piston moves up, pushing the waste gases out through the exhaust system. At the top of the stroke, the exhaust valve closes, the inlet valve opens and the piston moves down for the next induction stroke.

During the cycle, mechanical energy is extracted from a high-temperature gas that has a lot of **inertial energy**. The lower-temperature gas that leaves the engine has a lower internal energy.

-🔸- **Jet engine, Rocket**

FREEZING POINT

Freezing point is the temperature at which a solid turns to liquid, or a liquid turns to solid.

The freezing point of a **liquid** will be lowered by impurities. The freezing point (or melting point) of a **solid** is used by chemists as a check for purity. If salt is added to water, the freezing point is lowered, which is why we put salt on our roads and paths in the winter. We hope that the air/road temperature does not reach the new lower freezing point and the mixture remains liquid.

-🔸- **Boiling point, Kinetic theory**

FREQUENCY

Frequency is the number of times that a particular event occurs in 1 second. A frequency of one event per second is called 1 **hertz** (1 Hz).

We usually come across this when dealing with **waves,** so frequency is the number of complete waves that pass a chosen point in 1 s. If a radio wave has a frequency of 98.4 MHz, 98,400,000 complete waves arrive at the radio aerial each second.

If you are listening to a sound, the frequency of the wave decides the pitch of the note that you hear. A greater frequency is heard as a higher pitch. A note of middle C is 256 Hz, and each octave higher doubles the frequency.

The colour of light is also decided by the frequency of the light waves. Red is a lower frequency than blue, and there is a steady change through that part of the **electromagnetic spectrum** that is visible.

> Remember: A higher frequency will also mean a shorter wavelength because the two are connected by the **wave equation**.

Sometimes we also measure the *periodic time,* which is the time taken for one complete vibration. This is connected to frequency by the equation

$$\text{periodic time} = \frac{1}{\text{frequency}}$$

Measuring the periodic time may sometimes be easier than measuring the frequency, and you can quickly change from one to the other.

-🔸- **Amplitude, Electromagnetic spectrum, Wave, Wavelength**

FRICTION

When two surfaces slide over each other there will be a **force** acting against the motion, called friction. The size of the force will depend on the material of both surfaces and on how 'rough' each one is.

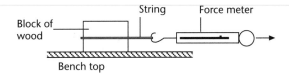

Fasten a force meter to a wooden block and use it to pull the block across the bench (see figure). At first the force increases, but the friction force is always equal and opposite to it and the block does not move. Note the largest value of the force when the block just begins to move. This is the maximum value of the 'static friction' for the two surfaces. Once the block is moving, the force needed to maintain a constant speed is smaller than the force of static friction and is called the force of 'dynamic friction'. You can investigate the frictional force between lots of surfaces using this apparatus. You should be able to show that size of the force depends on the perpendicular force between the surfaces (weight if you are always pulling the block horizontally) but does not vary much with surface area if the surfaces are dry. The value of frictional forces can be changed by various lubricants (scope for another investigation!).

Friction is often thought to be a nuisance in machines where heat is produced at sliding surfaces, because

heat = energy transformed = work done
    = frictional force × distance moved

but it is also necessary – we could not walk if there were no frictional force between the soles of our feet and the ground, and tyres work because of the frictional force between the tyre and the road. We also use frictional forces in most braking systems.

When working out the size of frictional forces you should remember that forces are in **equilibrium** (balanced, equal and opposite) when no acceleration is produced.

> Remember: A cyclist pedalling along a level road at a constant speed must have balanced forces. The force from the pedals is being balanced by a counter force from friction and air resistance.

-✛- *Force*

FULL-WAVE RECTIFIER

This is sometimes called a 'bridge rectifier' and uses four **diodes** to convert **alternating current** to **direct current**.

An **oscilloscope** connected across the a.c. input shows that the input is a full wave. If the oscilloscope is connected across the output at XY (see figure) it shows that the output is d.c. No matter which the original a.c. sends current, the diodes make it go through R_L in the same direction. R_L can be a resistor to demonstrate the circuit but, in a real circuit, would be the device that needs the d.c. The output may not be constant (smooth) enough for some purposes, and a large **capacitor** connected across the output at AB will help to make the output more like that of a battery.

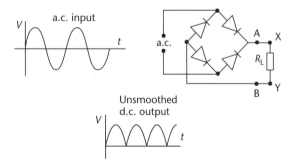

The action of the 'smoothing capacitor' in the figure below is explained under **half-wave rectifier**. In this case, the 'ripple' on the final output will be less than that of the half-wave rectifier because the

capacitor is recharged twice as often. Bridge rectifiers of this type are supplied in one integrated circuit containing the four diodes, so that you do not have to solder in the four individual components.

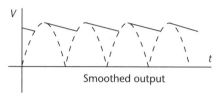

Smoothed output

-✛- *Capacitor, Diode, Half-wave rectifier*

FUSE

A fuse is a short length of wire that is fitted into a circuit in **series** with the supply. Its size and material are carefully calculated so that it will melt if a **current** flows through it that is larger than expected. This will leave a gap in the circuit that switches the current off and avoids overloading the supply cable and damaging it by heat. It may, incidentally, prevent further damage to the appliance.

> Remember: A fuse is there to prevent too much current going through the cable. This could overheat and damage the insulation and cause a fire. The fuse is NOT there to prevent electric shocks (see **Earth**).

To make them easy to replace, most modern fuses are in a cartridge form (see figure). The ends of the wire are connected to metal caps for easy connection to the circuit, and the wire is enclosed in a plastic or glass tube. Obviously, the smallest value of fuse that will carry the normal current of the circuit should be used. For example, a 'mains' table lamp with a 60 W bulb takes $60/240 = ¼$ A and does *not* need a 13 A fuse in its plug! The fuses that can be fitted into 'mains' plugs are 13 A, 5 A, 3 A.

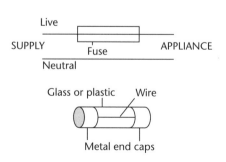

New mains installations may have miniature circuit breakers in place of fuses. These turn off the circuit when the current rises above a certain level.

They can work much faster than a fuse and can be reset by a simple switch instead of having to replace the fuse.

CHECKPOINT

Fred buys a small electric heater for his bedroom. It is rated at 1,000 W 230 V. What fuse should he have in the mains plug?

-✛- *Current, Domestic electricity, Plug, Power*

FUSION REACTION

In a fusion reaction, the nuclei of small atoms such as hydrogen and helium join together to make bigger nuclei, which need less energy to bind them together. The unwanted energy is released as thermal (heat) energy. This is the reaction that powers stars and happens in the very high temperatures and pressures near the star centre.

The reaction probably happens in three stages:

$$^1H + {}^1H \rightarrow {}^2H + e^+ + \nu$$
$$^2H + {}^1H \rightarrow {}^3He + \gamma$$
$$^3He + {}^3He \rightarrow {}^4He + {}^1H + {}^1H$$

Overall, the protons (hydrogen nuclei) are turned into helium nuclei, and particles called neutrinos and positrons are released along with gamma radiation and energy. The nuclei are positive charged and repel each other. To get them to join needs the equivalent of a temperature of millions of degrees.

-✛- *Fission, Star*

GALAXY

A galaxy is a cluster of hundreds of billions of stars.

- Our star, the Sun, belongs to a galaxy called the Milky Way. The Milky Way is a huge flat rotating spiral of stars. The Sun is a star about two-thirds of the way out from the centre. If you look up into the night sky you can see the Milky Way as a bright band of stars across the sky. You see it like this because you are looking at the flat spiral edge on.

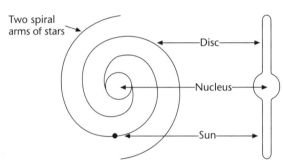

A spiral galaxy

- There are about a hundred billion galaxies in the universe.

- Galaxies are very large. Ours, the Milky Way is at least 100,000 **light years** across.

- Galaxies are often close to each other and form *clusters*. The Milky Way is part of a cluster called the '*local group*'. The other big galaxy in the group is called the Andromeda Galaxy.

-**∴** *Light year, Star, Sun*

GALVANOMETER

A galvanometer is a meter that will measure or detect small electric currents.

-**∴** *Ammeter, Electric motor effect*

GAMMA RADIATION (γ)

This is short-wavelength, high-frequency electromagnetic radiation emitted from the nucleus of atoms as part of radioactive decay and is one of the three common types of radioactivity (the others are alpha and beta).

- Gamma radiation occurs as short bursts of waves called *photons*, which have no mass and no electric charge. It is not deflected by either electric or magnetic fields.

- Gamma radiation has a lot of energy and can damage or kill cells when it hits them.

- Gamma radiation is very penetrating and will go through most materials, but damage is caused only when it collides directly with an atom, releasing a lot of energy.

- Great care must be taken to keep gamma sources in their proper containers and as far as possible from all parts of the body. Cells and body tissue are damaged by short-term burns that can kill cells and longer-term cancers caused by damage to the DNA of the cell.

- A few centimetre of lead will reduce the intensity by half, but doubling the thickness only halves it again, so a lot of 'shielding' may be needed.

- Distance is the best defence, because the radiation will be spread out over a larger area as it gets further from the source.

- Cobalt-60 is a source that emits only gamma radiation, but many other sources emit gamma after emitting alpha or beta radiation.

- Uses include sterilizing food and medical instruments (the radiation kills the bacteria) and treating cancer.

- This radiation can be detected by a Geiger counter, which picks up the ionization caused by the gamma radiation hitting atoms in the detector tube. It is important to realize that most of the radiation is not being detected, because it goes straight through the tube without hitting anything.

- The nucleus that emits the gamma photon will not change in mass or charge (it still has the same atomic number and atomic mass number), but it will have less energy than before.

-**∴** *Alpha particles, Beta particles, Electromagnetic spectrum, Half life, Radioactive decay*

GASES

Gases have no fixed structure, and their particles are free to move in any direction. Since there are large spaces between the particles, and the particles are moving rapidly in random directions, a gas is a fluid that can completely fill any shape of container and will diffuse rapidly. This rapid random motion can be shown in **Brownian motion**. Increasing the internal energy of the gas by putting in heat will make the particles go faster and the temperature will rise without further changes in state.

As the particles move they collide with each other and with the walls, but they do not slow down, because the collisions are totally elastic. The frequent collisions with the walls of the container produce a force on the walls, so a gas exerts a **pressure**, which depends on the energy of the particles when they bump into the walls.

If the gas is compressed into a smaller space, the particles hit the walls more often and the pressure increases. When you use a bicycle pump it pushes the gas into a smaller space until the pressure is large enough to open the valve and let the air into the tyre (see **Boyle's law**).

If the gas gets warmer the particles go faster, hit the walls more often, and the pressure rises. This will cause the gas to expand if it can (**Charles' law**) or the pressure will rise at constant volume (**pressure law**). This is why it is dangerous to throw unwanted tins or aerosols into a fire – the pressure will increase until the can explodes.

These pressure, volume and temperature effects are combined into the **general gas equation**.

> *Remember: Gas pressure is caused by the rapidly moving molecules hitting the walls of the container and not by a downward weight. This is also true of air pressure. The collisions and therefore the pressure act in all directions.*

⇌ **Kinetic theory, Liquids, Solids**

GEIGER COUNTER

This is a device for counting the particles emitted from a radioactive source. It consists of a power supply, an electronic counter or rate-meter and a detector. The detector was invented by Geiger and Müller and works by detecting the **ions** left in the path of particles passing through the detector.

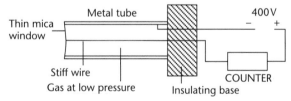

Geiger–Müller tube

The front mica window is very thin and will allow α, β and γ to pass through. The **alpha, beta** and some of the **gamma** protons will create ions as they pass through the gas in the tube. The **electrons** produced are accelerated towards the centre wire and go fast enough to hit atoms and create more ions. These electrons are also accelerated, and the process rapidly produces a lot of electrons that reach the wire and flow around the circuit as a short burst of current. This pulse is counted electronically as one particle. In some machines the pulse of current is amplified and sent to a loudspeaker, where it produces a 'click'. In other machines, the display shows the average count every minute or second instead of just the total (it is then a rate-meter rather than a simple

counter). Notice that the detector cannot count all the gamma photons that pass through the tube, as many of them go through without hitting anything and causing ions.

More sensitive and accurate detectors are now available, but the G–M tube is still in common use.

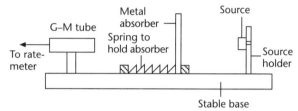

The Geiger counter is often used in the laboratory to find the penetrating power of a radioactive source, as this is one simple way of deciding which types of radiation are being emitted (see figure above). The background count is measured before the source is removed from its box and is measured again at the end of the experiment. The average value is subtracted from all the readings in the experiment. Various materials are put between the source and the detector and the count rate noted in each case, making sure that the distance from source to detector stays the same. For accurate experiments on a particular source you can use absorbers of exact thickness of aluminium or lead and plot a graph of counts/second against thickness of metal. To identify the type of radiation it will be enough to remember that alpha will be stopped by a sheet of paper, beta will penetrate the paper but be stopped by a few mm of aluminium, but gamma will be reduced only by thick sheets of lead, some always getting through. Moving the detector away from the source will enable the range of the radiation in air to be measured.

⇌ **Alpha particles, Beta particles, Gamma radiation, Half life, Radioactive decay**

GENERAL GAS EQUATION

This law enables you to work out what happens to the pressure, volume or temperature of a gas. It combines **Boyle's law, Charles' law** and the **pressure law**.

$$\frac{P_1 V_1}{T_1} = \frac{P_2 V_2}{T_2}$$

where P_1 = first pressure, V_1 = first volume, T_1 = first temperature, P_2 = second pressure, V_2 = second volume, T_2 = second temperature.

The temperature *must* be in kelvin (see **Temperature scale**). To convert from °C to K add 273.

Worked example

A gas has a volume of 200 ml at 5 atmos. pressure and 327 °C. What will its volume be at 1 atmos. pressure and 27 °C?

$$T_1 = 327 + 273 = 600 \text{ K}$$
$$T_2 = 27 + 273 = 300 \text{ K}$$

Using the gas equation

$$\frac{P_1 V_1}{T_1} = \frac{P_2 V_2}{T_2}$$

$$\frac{5 \times 200}{600} = \frac{1 \times V_2}{300}$$

$$V_2 = \frac{5 \times 200 \times 300}{600}$$

$$= 500 \text{ ml}$$

Do not forget to change any temperature in °C into K. You will see what the scale is all about if you look up **Temperature scale** and why it is needed here if you look up **Kinetic theory**.

-⧊- **Boyle's law, Charles' law, Kinetic theory, Pressure law, Temperature scale**

GEOTHERMAL ENERGY

Geothermal energy is used by pumping water under pressure down to hot rocks below the Earth's surface. The water returns heated and the heat energy can be used as an alternative resource. The original source of the energy is radioactive decay within the Earth's core. The process has been tested in a number of areas, including Cornwall, but few good sites have been identified in this country. It has been developed in other countries, such as New Zealand.

-⧊- **Sources of energy**

GRAPH

In many experiments you make measurements of two quantities to see whether they are connected in some way. It is usually difficult to see such a connection from a table of numbers, and plotting a graph makes a connection easier to find. If the points are on a smooth curve, there is a law connecting the two quantities, but it *might* be complex. If the points lie on a straight line through the origin, the two quantities are directly proportional. If the points are randomly scattered, there is probably no connection between the two quantities.

When you design such an experiment make sure that only the two quantities that you measure are allowed to change. Change one of the quantities (called the 'independent variable') in equal steps and record the second quantity (the 'dependent variable') as you go. Repeat all values if possible to check for **rogue results**. Plot the independent variable on the horizontal axis.

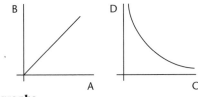

Typical graphs

The straight-line graph shows that A is directly proportional to B. Double one and you double the other.

The curve shows a graph where C is inversely proportional to D. Double one and you halve the other.

> *Remember: An inverse proportion does NOT give a straight-line graph. You can turn it into a straight line by plotting the reciprocal of one variable against the other, i.e. 1/C against D.*

-⧊- **Rogue results**

GRAVITATIONAL FIELD

Any two bodies attract each other with a **force** that depends on their masses and the square of their distance apart. This force is called the force of gravity. This force acts through space and, like magnetic and electric forces, does not need a material to travel in but decreases quite rapidly with increasing distance. The force has a significant size only when at least one of the bodies is very massive.

The *gravitational field strength* is the force acting on a mass of 1 kg at that place in the field.

The gravitational field strength at the surface of the Earth is 9.81 N/kg, but it does vary a little depending on your exact position. Most examinations accept 10 N/kg as accurate enough – and it does simplify calculations!

-⧊- **Force, Weight**

GRAVITATIONAL POTENTIAL ENERGY

Gravitational potential energy is the energy that the body has because of its height.

You can work out the change in gravitational potential **energy** by calculating the **work** that would be done to get to that position (see figure).

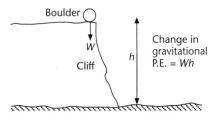

$$\text{work} = \text{force} \times \frac{\text{distance}}{\text{moved}}$$

$$\text{change in gravitational P.E.} = \text{weight} \times \frac{\text{change in}}{\text{height}}$$

$$= W \times h$$

The energy will be in joules if the weight is in N and the height is in m. Remember that the weight is a **force** in newtons – you *cannot* use mass (kg) in this equation. If you are given the mass you will need to find the **weight** first, using $W = mg$. Then:

$$\text{change in gravitational P.E.} = mgh.$$

Worked example

A girl has a weight of 600 N. How much gravitational potential energy will she gain if she climbs to the top of some steps that are 5 m high?

$$
\begin{aligned}
\text{change in gravitational P.E.} &= \text{weight} \times \text{change in height} \\
&= 600 \times 5 \\
&= 3{,}000 \, \text{J}
\end{aligned}
$$

CHECKPOINT

How much gravitational potential energy does 1 m³ of water have at the top of Niagara Falls? (The Falls are 45 m high, the density of water is 1,000 kg/m³ $g = 10$ N/kg.)

✤ *Energy, Potential energy, Weight*

GRAVITY

✤ *Gravitational field, Weight*

GREENHOUSE EFFECT

This phenomenon results in the Earth being warmer than it otherwise would be, but small increases in the future may change our climate. It is caused by gases like carbon dioxide in the atmosphere.

The Earth absorbs short-wavelength infrared light and other radiation from the Sun. This is balanced by the Earth radiating long-wavelength infrared so that its temperature stays the same. The Earth radiates longer waves than the Sun because it is cooler. Greenhouse gases allow the shorter waves in but absorbs the longer ones going out. Some of this energy is then radiated back towards the Earth. Extra energy is therefore trapped, and the Earth becomes warmer as a result. This can have a great effect on climate patterns, and parts of the polar ice caps could melt in the future.

The main greenhouse gases are carbon dioxide, water vapour and methane, produced by burning fossil fuels.

✤ *Electromagnetic spectrum*

HALF LIFE

A half life is the time taken for half of the atoms in a sample of a radioactive nuclide to decay.

The atoms of some **nuclides** are totally stable and do not decay at all. Others have a nucleus that is not quite stable and will emit either **alpha, beta** or **gamma** radiation. The more unstable the nucleus the shorter the half life will be. Note that it is impossible to tell when one particular atom will decay, but we can measure what will happen when large numbers of the atoms are present. Half lives can vary from fractions of a second to millions of years.

Worked example
A radioactive material is found to have a half life of 4 days. If 60 g is put into a container, how much of the radioisotope will be left after 12 days?

$$\begin{array}{ccccc} 4\ days & & 4\ days & & 4\ days \\ 60\,g & \rightarrow & 30\,g & \rightarrow & 15\,g & \rightarrow & 7.5\,g \end{array}$$

The diagram shows that 12 days is three half lives and that the remaining material has a mass of 7.5 g.

Remember: The other 52.5 g will not have disappeared. It will now be a different nuclide, so that the total mass will still be 60 g if you put the sample on a balance. If you were trying to measure half life this way you would have to do a chemical analysis of the sample at each stage and, since this takes quite a long time, the method would work only for long half lives.

CHECKPOINT

A particular nuclide decays to produce another nuclide that is not radioactive and has a half life of 20 min. The count rate from a sample of the nuclide is 1,600 counts/second. What will the count rate be after 2 hours?

Finding half lives from graphs

If you are given a table of data, the half life may be obvious from the pattern of the numbers, but it is more likely that you will need to plot a graph to make what is happening clearer (see figure). Plot the time on the horizontal axis and the mass or count rate on the vertical axis, using the largest scales

possible. This should produce a smooth curve, and any **rogue results** should be easy to spot.

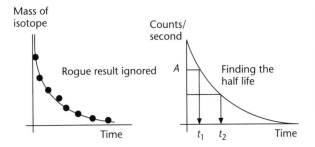

After fitting in the smooth curve, choose any value (*A*) on the mass (or count rate) axis and find the time (t_1) when it happens. Now halve the value of *A* and find the time (t_2) when that happens. The time from t_1 to t_2 is the half life. Repeat the process with different values of *A* and you should always get the same answer.

Experiments to find a half life

The process is aimed at getting a set of data so that you can plot a graph and carry out the process in the last section. Finding the mass is difficult (see the worked example above) and could be dangerous because of the risks of using radioactive powders and liquids (see **Radioactive decay**). The following method illustrates the method that can be used if the 'daughter' nuclide is not radioactive.

The 'parent' nuclide in this case is a radioactive **isotope** of a gas called radon. ^{220}Rn is given off by thorium hydroxide, which can be kept as a powder in a sealed plastic 'squeezy' bottle. A sample of the gas can be introduced into a chamber that is part of a closed circuit, as shown below, by squeezing the bottle and then closing the clips on the rubber tubing. The chamber also contains the detector tube of a **Geiger counter**. The count rate is noted at one-minute intervals so that a graph of count rate against time can be plotted. The half life can then be found as described in the previous section. Since the half life is fairly short (52 s) and the daughter nuclide is safe (not radioactive), the chamber can be opened without danger after a few hours.

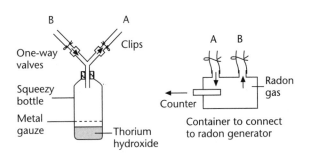

Half lives and safety

In most cases, the safest work will be carried out with the smallest possible sample of the shortest half life nuclide that will perform the task. This will reduce the health risk in the case of an accident by releasing the smallest possible quantity of radiation. A small sample of a short half life nuclide will be just as active and dangerous as a much larger sample of a long half life nuclide, and radioactive materials are usually sold in amounts that depend on their activity and not simply their mass.

✦ *Alpha particles, Beta particles, Gamma radiation, Radioactive decay*

HALF-WAVE RECTIFIER

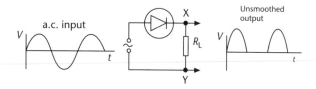

This circuit uses a diode to turn **alternating current** into **direct current**. The diode lets current through in only one direction.

An **oscilloscope** connected across the a.c. input shows that the input is a full wave. If the oscilloscope is connected across the output at XY it shows that the output is d.c. but is half the original wave, so that it is in short pulses with a space between them. R_L can be a resistor to demonstrate the circuit but, in a real circuit, would be the device that needs the d.c. The output may not be constant (smooth) enough for some purposes, and a large **capacitor** connected across the output will help to make the output more like that of a battery (see following figure).

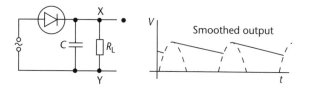

The capacitor C acts as a reservoir of charge. Each time the voltage rises the capacitor is recharged to the full voltage. As the voltage of the supply falls, the capacitor can still supply current to R_L, so the voltage is kept fairly steady.

The voltage on the capacitor does fall slightly as charge is taken from it, but this 'ripple' can be kept small if the capacitor is big enough. If a **full-wave rectifier** is used the capacitor is recharged twice as often and the ripple is correspondingly smaller.

HEATING EFFECT OF ELECTRIC CURRENT

If you refer to the entry on **Potential difference**, you will find that 1 joule of energy is released for each coulomb of electric charge that flows through a p.d. of 1 volt. This then means that the energy released in any part of a circuit is given by:

$$\text{energy changed} = VIt$$

where V = p.d. in volts, I = current in amps, t = time in seconds, the energy is in joules.

If the energy is being released in a **resistance** in the circuit it will appear as heat, the component that has the resistance getting hotter. This happens in the filaments of light bulbs and the elements of electric fires.

It is also true that the power is the energy changed per second, and putting $t = 1$ into the equation gives:

$$\text{power} = VI$$

which is a useful way of finding the rate at which you are using electrical energy in a circuit. The unit for **power** is the watt (W) and 1,000 W = 1 kW.

Sometimes you do not know the p.d. and the current, but you do know the **current** and the resistance instead. If you remember **Ohm's law** and that:

$$V = IR$$

you can put IR in place of V in the energy equation and get

$$\text{energy changed} = I^2Rt$$

where I is the current in amps, R is the resistance in Ω and t is the time in seconds.

> Remember: This shows you that doubling the current through a particular resistance will produce four times as much heat, not twice as much, because it depends on current squared!

The power is the heat energy per second, so we also know that:

$$\text{power} = I^2R$$

Sometimes these units are too small for particular use. This is especially true when we know we are paying for electrical energy, and we need a larger unit, the kilowatt-hour (see **Payment for electrical energy**).

Worked example
A bulb for use in a ray box is marked 12 V 24 W.
1. What does this marking mean?
2. How much energy will it use if it is switched on for 5 mins?
3. What current does it take when connected to a 12 V battery?
4. What is the resistance of the bulb?

1. The bulb should be connected to a 12 V supply for it to operate at its correct brightness,. When properly connected it will convert energy at a rate of 24 W, or 24 joules per second.

2. Energy changed = power × time
$$= 24 \times 5 \times 60$$
$$= 7{,}200 \text{ J or } 7.2 \text{ kJ}$$

3. Power $= VI$
$$24 = 12 \times I$$
$$I = \frac{24}{12} = 2 \text{ A}$$

4. We know both p.d. and current, so we can use Ohm's law:

resistance $= \dfrac{V}{I}$
$$= \frac{12}{2} = 6 \,\Omega$$

CHECKPOINT

How fast is heat energy being produced in a
100 Ω resistor that is carrying a current of 0.5 A?

-❖- **Current, Ohm's law, Payment for electrical energy, Potential difference, Power, Resistance**

HEAT TRANSFER

There are three main methods by which heat is transferred from one place to another. **Conduction** occurs in solids and liquids and is best in metals. **Convection** occurs in liquids and gases (fluids) and not at all in solids. Radiation is **infrared radiation** and transmits heat well through gases but is absorbed by solids.

-❖- **Conduction of heat, Convection, Infrared radiation**

HERTZ (Hz)

-❖- **Frequency**

HOOKE'S LAW

The deformation of a body is directly proportional to the **force** producing it. The law applies only to *elastic* materials. These materials make a restoring force that tries to return the body to its original shape. Springs, wooden beams, girders and rubber bands are good examples of elastic materials. *Plastic* materials such as plasticene or putty can be moulded into a new shape and have not got the 'springiness' of elastic materials.

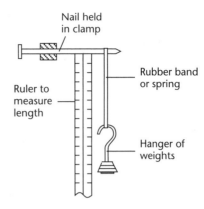

Nail held in clamp

Ruler to measure length

Rubber band or spring

Hanger of weights

Many different materials will stretch when a force is applied and return to their original length when it is removed. Examples are thin copper wire, rubber bands and springs. Take one of these, hang it from a clamp on a stand and fit a weight hanger on the other end (see figure). Measure its original length. Hang a weight from its end and record the weight and the new length in a table. In the next column of the table work out the *extension*, which is the total amount that the material has stretched. Do this several more times, checking the weight and the extension each time. You may be able to find a more direct way of measuring the extension with your particular apparatus, but make sure that it is the *total* distance stretched. Plot a graph of load (the weight added in N) on the *x*-axis against the extension on the *y*-axis. The graph will start off straight but eventually bend if you add a lot of load – you can try to do this if you are stretching wire, but your teacher will be very annoyed if you permanently damage an expensive spring! The straight part of the graph shows the material obeying Hooke's law, because the extension is proportional to the load.

The point where the graph stops being straight is the *elastic limit*, and the material stops obeying Hooke's law at this point. That is, the elastic limit is the point at which the deformation stops being proportional to the load. The material will begin to extend more than before for the same extra load. If the load is removed, the material does not go back to its original length and has a 'permanent set'.

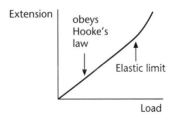

Extension

obeys Hooke's law

Elastic limit

Load

A **newton meter** uses this law to measure forces. The more the spring extends the greater the force applied. A strip of metal attached to the end of the spring carries the point, which reaches the scale through a slot in the metal case. Care should be taken not to exceed the elastic limit – start with a stronger meter and change to a smaller one when you are sure it is right to do so.

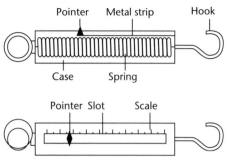

A newton meter

Many other systems obey Hooke's law, including beams, and you can show the principle using a metre rule. Clamp one end of the rule to your bench and hang weights on the other end so that it bends. You should be able to show that the distance that the end moves downwards is proportional to the load. Incidentally, the top of the ruler is then under **tension** and the bottom of the ruler is under **compression**. The centre of the beam is neither and can be made thinner, as it does not need to be as strong – many steel girders have an I-shaped cross-section for this reason.

Worked example

A climbing rope stretches by 100 mm when a girl of weight 400 N has a fall. If a man of weight 700 N fell the same distance on the same rope how much would it stretch?

400 N produces an extension of 100 mm

1 N produces an extension of $\dfrac{100}{400}$ mm

700 N produces an extension of $\dfrac{100 \times 700}{400}$ mm

The rope stretches by 175 mm, provided that the elastic limit is not exceeded. The rope needs to be chosen with care so that its elastic limit is not exceeded and it does not extend too much or too little.

> ### CHECKPOINT
>
> A spring is 250 mm long and can be stretched by 50 mm if a 200 N load is applied. How long will the spring become if a load of 100 N is applied to it?

-✛- Force

HYDROELECTRIC POWER

Hydroelectric power is used in many countries that have rivers large enough to provide the energy required. In this country, most rivers do not fall through enough height or have a large enough rate of flow to produce large quantities of energy, but many are used on a smaller scale. The gravitational potential energy of the water is converted into kinetic energy as it falls and then into electrical energy, by turbines (see **Energy**). The original source of the energy is the Sun, which drives the water cycle.

-✛- Gravitational potential energy, Sources of energy

IDEAL GAS EQUATION

-**-** *General gas equation*

INERTIA

-**-** *Mass*

INFRARED RADIATION

These *waves* are part of the *electromagnetic spectrum* and are emitted by all hot bodies. The wavelengths that are emitted get shorter as the temperature increases. The waves are easily absorbed and their energy becomes heat.

An electric fire works by having a red-hot element that sends out a lot of infrared radiation. This can be reflected forward by the metal reflector behind the element and will turn back into heat when it is absorbed by walls, furniture, people, etc. These will in turn radiate heat at a longer wavelength.

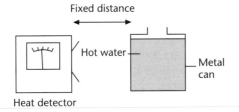

The apparatus in the figure above shows the effect of the type of surface on the radiation that is emitted. The metal can is filled with hot water and has four sides that are matt black, shiny black, white and polished metal. These will all be at the same temperature because the can is a good *conductor*. The radiation (infrared) detector is pointed at each surface in turn and from the same distance each time. It shows that the surfaces all radiate heat, but that the dull black is best, followed by the shiny black. The polished surface emits very little radiation. Marathon runners may be protected by metal foil blankets after the race to prevent them losing more body heat. We use matt black when we paint engines, or transistor heat sinks, so that they radiate heat as quickly as possible. An air-cooled engine such as a motor-cycle engine will have its cylinders black on the outside to get rid of heat as quickly as possible. Coating the surface with shiny chromium plate can cause damage through overheating if the

engine is not designed for it. The engine will also have a lot of fins to increase the surface area – this will be able to radiate more heat as well as being in contact with more cool air when the machine is moving.

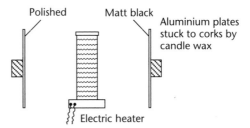

The effect of the surface on the absorption of heat can also be checked using the simple apparatus in the figure above. The two metal plates are identical pieces of aluminium with a different surface treatment, and are the same distance from the heater. The surface that absorbs the heat best will fall off its cork first as the wax melts. This can be done with a variety of surfaces. The matt black surface is the best absorber. Polished surfaces act as a mirror and reflect the waves so that they remain cool; i.e. the best absorbers are also the best radiators. Buildings in hot countries are often painted white and have their roofs painted with silver paint so that they absorb less heat and stay cool in the hot sun. Satellites often have a protective layer of silver foil to reflect away radiation from the Sun and prevent overheating.

-**-** *Conduction of heat, Convection, Vacuum flask*

INSULATOR

An insulator is a material through which it is difficult to pass heat or an electric *current*.

Non-metals will usually be insulators because they do not have 'free' *electrons* to carry the energy or electric charge through the material. Plastics are good insulators because they do not have these free electrons and are not made of *ions* (they are covalently bonded). Their *molecules* are also very large and interwoven and cannot transmit heat very easily by vibration.

Gases are poor conductors of heat because of the spaces between the particles, which stop the *kinetic energy* being passed from one particle to another. If the gas is kept in small pockets or thin layers it will not be able to transfer heat by *convection* and will produce a good insulator (see figure).

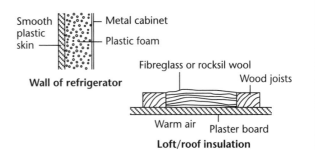

53

Examples are plastic foams, which set solid and contain pockets of carbon dioxide or air and are used in the walls of fridges and some types of cavity wall insulation. Layers of clothing trap layers of air between them, and two thin layers are often warmer than one thick one. Roof insulation works in the same way, preventing heat loss from the warm air, which rises to a layer under the ceiling as a part of the process of **convection**. Lagging on water pipes and the insulating jacket on hot-water cylinders also trap air in layers of plastic foam.

-✦- *Conduction of heat, Conductor, Convection*

INTEGRATED CIRCUIT (IC)

An integrated circuit is a miniature electronic circuit that has been built up by depositing and etching away layers of **semiconductor** to make a 'silicon chip'. An IC may contain hundreds of **transistors** and **resistors** on a very small area. This tiny circuit is then put into a plastic container, which protects it, and connections to it are made through metal 'pins'. A simple IC may have very few connections to it and look rather like a transistor, but most need many connections to carry the signals in and out.

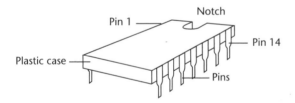

The usual shape is a rectangular black plastic package with a row of connections down each side (see figure). One end has a 'notch' in it so that you can tell which pin is which. This is called a **DIL** (dual in-line) chip. **Logic gates, op amps**, and microprocessor chips are all different types and sizes of IC. Many of these chips look exactly alike – you can tell what they do and what connects to what only by looking up the number on the top of the case. For example, a 74LS08 is a 14-pin **DIL** chip. Two of its pins connect to the supply voltage and it contains four AND gates, which each use two inputs and one output – fourteen connections in all.

-✦- *Logic gates, Transistor*

INTERFERENCE

When the paths of two **waves** of the same type cross they interfere. Their displacements are added together.

Sometimes the two waves arrive exactly together, so that both peaks push in the same direction at the same time, producing a bigger wave. The waves are 'in phase'. This is constructive interference.

Constructive interference

Sometimes the waves are pushing the medium in opposite directions, so that the waves cancel each other. If two opposing 'peaks' (i.e a peak and a 'trough') arrive at the same time, the waves are out of phase and equal amplitudes would exactly cancel. This is called destructive interference.

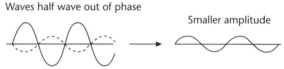

Destructive interference

In most places, the interference will be somewhere between these two extremes. The two waves will pass completely through the interference area, where they cross and then carry on as before.

-✦- *Ripple tank, Wave, Young's slits experiment*

INTERNAL ENERGY

The internal energy of a body is the sum of all the kinetic *and* potential energy of its particles. When energy is put into a body, its internal energy is increased by the same amount. If there is no change of state, the particles gain **kinetic energy** and we notice the faster molecules as a rise in temperature. If there is a change of state, the molecules gain **potential energy**, and we call the energy taken in **latent heat**.

-✦- *Energy, Latent heat, Specific heat capacity*

INVESTIGATION

See **Appendix Four** at the end of the book.

-✦- *Graph, Line of best fit*

ION

An ion is an **atom** that has had one or more **electrons** removed from, or added to, its energy levels. This will make it a charged particle.

This can happen during the formation of a chemical compound with ionic bonding. In this case, electrons are transferred from one atom to the other. For instance, in sodium chloride (common salt) the sodium atom gives one electron to the chlorine atom so that the sodium chloride is made up of sodium ions (Na^+) and chlorine atoms (Cl^-).

In physics, the ions are usually caused by an atom being hit by a fast-moving particle. This knocks one or more electrons off the atom, to form an ion. The fast particle may do this many times, losing a little energy each time until it stops. This process is called *ionization*. More ions can be created by a faster or more massive particle, because it carries more energy.

-♦- *Alpha particles, Atom, Beta particles, Gamma radiation, Radioactive decay*

ISOTOPE

Isotopes are **atoms** of the same element that have different numbers of neutrons in their nucleus.

This will mean that the isotopes have the same *proton number*, because they are atoms of the same element, but a different *nucleon number*. All elements have isotopes, but some are very rare. Oxygen is 99.76 per cent ^{16}O but 0.2 per cent of oxygen is ^{18}O. Chlorine is 75 per cent ^{35}Cl and 25 per cent ^{37}Cl, so its average atomic mass is 35.5. Most of these

isotopes are stable and quite normal. Some have a nucleus that is not quite stable because it has the wrong balance of protons and neutrons, and these will be **radioactive.** ^{12}C, ^{13}C and ^{14}C are all isotopes of carbon, but only ^{14}C is radioactive.

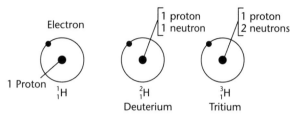

Hydrogen isotopes

Hydrogen has three isotopes, ^1H, ^2H and ^3H, as shown in the figure. Deuterium and tritium join together in the *fusion reaction* that powers the Sun, and a lot of research is being done to try to make this work on a small scale for use in power stations.

> *Remember: Not all isotopes are radioactive*

-♦- *Atom*

JET ENGINE

The jet engine works by using **Newton's third law**. It takes air in at the front of the engine and compresses it before burning it with fuel in a combustion chamber. The fuel is kerosene (like paraffin), and a hot expanding gas is produced, with fast-moving molecules. This passes through a turbine to provide the energy for the compressor at the front of the engine and then leaves the engine through the jet at the back (see figure). If a stream of high-energy particles leaves the rear of the engine with a lot of momentum, the engine is pushed forward with the same rate of change of momentum (force), in accordance with Newton's third law.

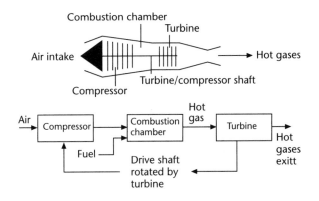

Remember: The jet cannot operate outside the atmosphere, because it uses oxygen from the air to burn with its fuel.

✦ **Newton's third law, Rocket**

JOULE

One joule of **work** is done when a **force** of one **newton** moves an object through a distance of one metre. The symbol for a joule is J.

✦ **Energy**

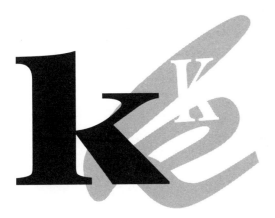

KELVIN

This is a temperature scale that starts from absolute zero. The unit is the *kelvin* and is the same size as 1 °C.

✦ **Absolute zero, Kinetic theory, Temperature scale**

KINETIC ENERGY

Kinetic energy is energy that a body has because it is moving. The **energy** will depend on the **mass** of the body and its **velocity** (see figure).

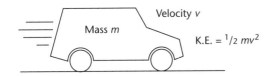

Velocity v

Mass m

K.E. = $^{1}/_{2}\,mv^2$

kinetic energy = ½ × mass × (velocity^2)

$$\text{K.E.} = \frac{m \times v^2}{2} = \tfrac{1}{2}mv^2$$

The energy will be in joules if the mass is in kg and the velocity is in m/s.

Even small masses can have a lot of energy if their velocity is large – a small bullet can travel rapidly!

Worked example
A car has a mass of 1,000 kg. What is its kinetic energy at

1) 10 m/s   2) 20 m/s?

1) K.E. = ½ × m × v^2
    = ½ × 1,000 × 10^2 = 50,000 J
2) K.E. = ½ × m × v^2
    = ½ × 1,000 × 20^2 = 200,000 J

CHECKPOINT

Write down the formula that connects kinetic energy, mass and velocity. A sprinter covers 100 m in 10 s. His mass is 75 kg. What is his average kinetic energy?

✦ **Energy, Mass, Velocity**

KINETIC THEORY

Kinetic theory is a theory that explains the behaviour of **solids, liquids** and **gases** by looking at the way that their particles behave. Kinetic means 'moving' and the theory assumes that all materials are made from very tiny moving particles. Depending on the material, these materials may be **atoms, molecules** or **ions**.

When energy is put in, the particles move more quickly.

In solids, the particles are held in their own place in the structure and can only vibrate. At the melting point, the energy being put in (**latent heat**) is used to break down the structure so that the particles can move more freely.

In a liquid, the particles are still close together and attract each other but can move about so the liquid flows. At the boiling point, the energy put in (an even larger latent heat) is used to separate the particles into a gas.

In a gas, the particles are much farther apart and move more quickly. The gas can therefore occupy any space that it is put into. A gas has pressure caused by its particles hitting the walls of its container.

✦ **Boiling point, Brownian motion, Change of state, Diffusion, Freezing point, Gases, General gas equation, Liquids, Solids, Temperature scales, Vapour**

LATENT HEAT

When a material changes state it takes in energy (even though the temperature stays the same), which is used to change the structure of the material. This energy is called latent heat. Latent means 'hidden' and the energy was given this name because scientists could not understand how heat could go in without the temperature rising. We now know that it is used to change the particle structure of the material (see **Kinetic theory**).

Each unit mass of a particular material will need a certain quantity of energy for the material to melt or to boil. This is called the *specific latent heat capacity of fusion* when the solid melts and the *specific latent heat of vaporization* when the liquid evaporates. Energy absorbed when the material melts is released when it freezes again. Energy absorbed to vaporize a liquid is released when it condenses again.

$$\text{heat of fusion} = \text{mass} \times \text{specific latent heat of fusion}$$
$$= mL_t$$

$$\text{heat of vaporization} = \text{mass} \times \text{specific latent heat of vaporization}$$
$$= mL_v$$

The specific latent heat will be in J/kg.

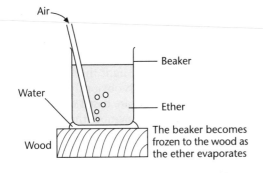

Heats of vaporization are usually much larger than the corresponding heats of fusion. It is much more dangerous to get a scald from steam than from the same mass of water at boiling point because it will also release this large latent heat. Spilling a liquid such as alcohol or butane or petrol on your hand can cause it to feel very cold as the liquid evaporates and removes the latent heat to do so from your hand. It is possible to use this effect as a mild anaesthetic before injections. We use the same effect to cool our bodies when we sweat.

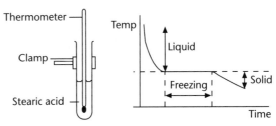

If a material is at a higher temperature than its surroundings it will lose heat to the surrounding materials. This will cause it to cool (see **Specific heat capacity**), except when it can give out latent heat instead. This shows up clearly if you plot a 'cooling curve' (see figure). You can obtain suitable data by using a water bath to melt some stearic acid in a boiling tube. Place a thermometer in it and take the temperature each minute as it cools. You will need to use the water bath again to extract the thermometer.

> *Remember: Latent heat has uses as well as causing problems. A liquid is turned into a vapour inside the cooling element of a refrigerator. This takes heat from the inside of the fridge. The vapour condenses back again in the radiator on the outside of the back of the fridge and releases the latent heat into the air.*

➤ **Kinetic theory, Refrigerator, Specific heat capacity**

LDR

LDR stands for light-dependent resistor. It is a **resistor** that will change its **resistance** depending on the brightness of the light that falls on its window (see figure). It will have a smaller resistance in brighter light. Different types from the different manufacturers vary in the amount of resistance produced, but one common type varies from a few hundred ohms in bright light to $100\,k\Omega$ in the dark.

There are many uses for these devices, including light meters for photographers and automatic switches for security lights outside homes and factories. Usually the rest of the electronic circuit will need an output voltage that changes with the light, and this is often produced by putting the LDR in series with a suitable resistor to form a **potential divider**, as shown in the second figure. In this circuit, the output voltage will increase in the dark because the LDR has a larger resistance and therefore gets a larger share of the voltage.

If you want a circuit that produces a smaller output voltage in the dark and larger in the light, you can simply interchange the two components so that the LDR is in the upper half of the potential divider. Whether you do this or not the circuit will still have a fixed output voltage for a particular brightness of light and this may not be suitable in

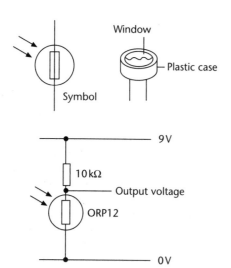

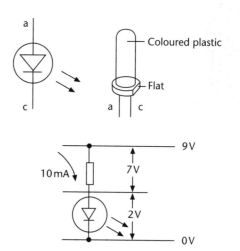

your design. The 'cure' is to use a variable resistor in place of the fixed one in the diagram. You will then be able to adjust the circuit to give the voltage required at a particular brightness.

This 'final' version with a variable resistor in series with the LDR forms a commonly used *light sensor* for triggering **logic gates** and **transistor switches**. If you have this sort of circuit in your school or college laboratory you will probably have used similar sensors and adjusted them to react to suitable light levels.

> Remember: These devices operate only at low voltages and will use transistor switches and relays to operate 'mains' equipment.

-✦- **Logic gates, Potential divider, Thermistor, Transistor switch**

LED

LED stands for light-emitting diode. It is a **semiconductor** diode that behaves in a similar way to an ordinary **diode** but gives out light when it is conducting current (see figure).

The LED conducts **current** and gives out light only when it is forward-biased and current goes in the correct direction. If a reverse bias is applied it must be kept small or the device will be damaged. The light produced is not very bright, but is very useful as a signal to indicate whether a circuit is on or off. The most common colour is red, but the devices are available in green and yellow and in a variety of different shapes.

When the LED is used it must have a suitable **resistor** connected in **series** that will limit the current passing through the diode. In most cases, a current of about 10 mA will make the diode bright enough, and the diode will be designed to work with about 2 V across it. This makes it possible to use **Ohm's law** to work out the value of the resistor required.

The circuit diagram in the second figure shows an

LED being used as a signal lamp to show when the 9 V supply is switched on. When it is on, the diode will have 2 V across it, leaving 7 V across the resistor. The current is 10 mA. Using Ohm's law on the resistor:

$$\text{resistance} = \frac{\text{p.d.}}{\text{current}}$$

$$= \frac{7}{0.01}$$

$$= 700\,\Omega$$

This does not need to be an exact value, and a resistor of 1 kΩ would do nicely.

-✦- **Diode, Ohm's law, Semiconductor, Seven-segment display**

LENGTH

The unit for length, the *metre*, is an important fundamental unit that we need to define exactly. It used to be the length of a platinum–iridium alloy bar that was kept in Paris, but this is no longer accurate enough. The metre is now defined as 1,650,763.73 wavelengths of the orange light that is emitted by [86]krypton. The definition is obviously important, because many other units are based on it, but you will not be expected to remember the numbers for examinations!

-✦- **Appendix One**

LENS

A lens is a piece of material, usually glass or hard plastic, that is specially shaped to refract light waves in a particular direction.

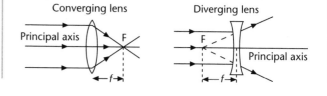

A converging lens will be thicker in the centre than around the edges. A diverging lens will be thinner at the centre than at the edges.

A lens has a point very close to its centre called the *optical centre*. The line through the optical centre at right angles to the lens is called the *principal axis*.

If light is sent into a converging lens parallel to the principal axis it will all pass through a point on the principal axis called the *principal focus, ('F' on the diagram)*. Light entering a diverging lens parallel to the principal axis will spread out as though it has come from a principal focus. In both cases, the distance from the optical centre of the lens to the principal focus is called the *focal length ('f' on the diagram)*. A converging lens will bring light (and heat) from the Sun to a focus – and also do the reverse by producing a parallel beam of light from a bulb at the focus.

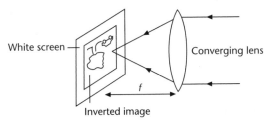

Light from a distant object is almost parallel, and we can use this to find the approximate focal length of a converging lens (see figure below). Move the lens backwards and forwards in front of a piece of plain white paper until the distant scene through a window is focused clearly on it. The distance from the lens to the paper is then the focal length.

You will notice when you do this that the image produced is *real* and *inverted*. This is the sort of image that is produced by the lens in a *camera*. You can produce different images from a converging lens, depending on its distance from the 'object'. The useful cases are shown in the figure as ray diagrams.

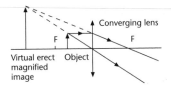

(a) Object nearer to lens than focus – used as magnifying glass

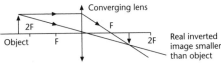

(b) Object between one and two focal lengths from lens used as projector lens

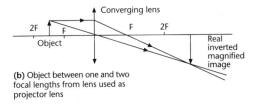

(c) Object more than two focal lengths from lens – the camera lens and eye lens work like this

Images from a converging lens

Notice that to locate the image we use the path taken by two rays leaving the same place on the object. One goes straight through the optical centre (this will happen provided that the lens is thin) and the other goes in parallel to the principal axis so that it comes out through the focus. You could draw this sort of diagram to scale if you had to find the exact position of an image, knowing the position of the object and the focal length of the lens. To make the diagrams easier to understand, the focus is called F and the point two focal lengths along the axis from the lens is called 2F.

An object closer to the lens than the focus produces an image that is erect (the same way up as the object), magnified and *virtual*. This is used in a magnifying glass. An object between one and two focal lengths from the lens produces an image that is inverted, magnified and *real*. This is used in a **projector** lens. An object further away than two focal lengths will produce an image that is inverted, diminished and *real*. The camera lens and the lens in the eye work in this way.

-**⊹**- *Camera, Projector, Real image, Virtual image*

LENZ'S LAW

When a **current** is induced in a circuit, it flows in such a direction as to oppose the change producing it.

-**⊹**- *Electromagnetic induction, Faraday's law*

LEVER

A lever is a **machine** that is designed to be a force multiplier, and it will always turn about a pivot.

The **force** that is put into a device is called the *effort*, and it will be used to move against another force, called the *load*. You can find the connection between the load and the effort by assuming that the forces just balance and then using the principle of **moments**.

Most levers belong to one of three types, as shown in the examples in the figure.

If the effort just balances the load on the crowbar, the principle of moments shows that

$$\text{load} \times L_2 = \text{effort} \times L_1$$

$$\frac{\text{load}}{\text{effort}} = \frac{L_1}{L_2}$$

Since L_1 is bigger than L_2 you have a very useful machine which is a *force multiplier*. In other words, the force that you put in (effort) is multiplied by quite a large number, so it can move a large load.

In the wheelbarrow, the load is between the pivot and the effort, but the distances still make the load larger than the effort (because the effort is farther away from the pivot than the load). It will be easier to move the wheelbarrow if the load is at the front, close to the wheel.

In the tongs or tweezers the load is at a greater distance from the pivot than the effort, so the load is less than the effort. This pair of levers is for convenience rather than for making a large force. Levers often come in pairs in simple tools.

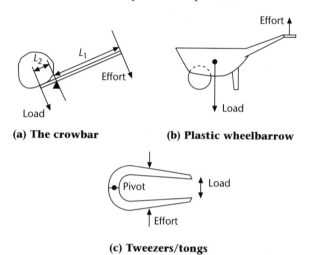

(a) The crowbar   **(b) Plastic wheelbarrow**

(c) Tweezers/tongs

When the muscles that we call the biceps pull on the forearm, the load in the hand is further away from the pivot (elbow) than the effort and the lever is of the third type. A pair of scissors or pliers will have two of the first type of lever, so that a small effort can act on quite a large load.

Many simple machines use levers. Some other examples are a spanner, the pedal of a cycle, a door handle, and nutcrackers.

-+- *Machine, Moment, Pulley*

LIFE CYCLE OF A STAR

A large cloud of dust in space, called a *nebula*, collapses as gravity pulls it together and shock waves from other events such as exploding stars (**supernovae**) clump some of the mass together. Really big clouds can form clusters of stars.

If this mass is small (less than about 1/10 of the Sun) it is too small to begin a **fusion reaction** and becomes a **brown dwarf**.

If the mass is bigger than this and less than about 80 times the Sun it will become a 'main sequence' star.

If the mass is bigger still it may form a very dense core and then explode as a supernova. Most of the mass will be scattered into space, but if the mass left behind is bigger than about three Suns it will collapse completely into a **neutron star** (**pulsar**). Larger stars may become so dense that not even light can escape, and the result is called a **black hole**.

In a main sequence star, the atoms in the centre are pushed closer and closer under the attraction of gravity, and a fusion reaction is started. The star then emits a lot of energy for billions of years during its stable period. The gravitational force pulling it together is balanced by thermal (heat)

forces. Bigger stars burn their nuclear fuel (hydrogen) more quickly and last for a shorter time. The Sun is about 4.6 billion years old and is about halfway through this period.

When the star runs out of hydrogen, it expands and cools to become a **red giant**. This may be hundreds of times bigger than the original star. During this stage, the star combines some of its helium to make bigger nuclei. Eventually, it contracts again and becomes a very dense **white dwarf**. After losing energy and cooling steadily for more millions of years, it will become a **black dwarf**.

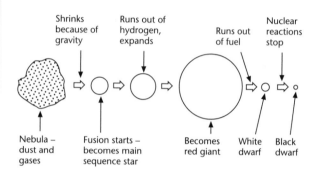

The life cycle of a main sequence star

-+- **Fusion reaction, Galaxy, Star**

LIGHT YEAR

A light year is the distance that light travels in a year. This huge distance is used to measure distances in astronomy, where kilometres would be far too small to have any meaning.

-+- **Galaxy, Universe**

LINEAR EXPANSIVITY

-+- **Expansion of solids**

LINE OF BEST FIT

All measurements have an error, even though you try hard to be as accurate as possible. When you plot the results of an experiment they will often be in a clear pattern, but the errors in the 'real' measurements will stop them being exactly on the line that you expect. Draw in a line of best fit that goes as close as possible to your points. This may be a straight line that does not go exactly through any of the points but is close to all of them. It may be a smooth curve, like the one for inverse proportion.

> *Remember: (1) If one of your points is clearly not part of the pattern it may be a rogue result and you should not move your line to go close to it.*
> *(2) Never join the points with a series of short straight lines.*
> *(3) Graphs that go from top left to bottom right are almost always curves and not straight.*

-ı̇- *Graph, Rogue result*

LIQUIDS

Liquids do not have the crystal structure of **solids**, and their particles are free to move about within the liquid. This will mean that the liquid does not have a fixed shape and that it will flow to take the shape of its container. It is a *fluid*. The particles are still very close together (there is often a change in volume on melting/freezing, but it is small compared with the total volume) and they still attract each other. This will make a liquid very difficult to compress into a smaller space, and **pressure** applied to any part of it will be found everywhere else.

Two liquids will be able to diffuse through each other, but the diffusion process is slow because there are no spaces between the particles. The process may be a little faster at higher temperatures, because the particles are moving faster. The movement is in random directions, as shown by **Brownian motion**.

If more energy is added, eventually some of the **molecules** at the surface of the liquid will be moving rapidly enough to escape from the attraction of the others and move into the space above, forming a **vapour**. Since these are the faster molecules, the remaining liquid has a lower average energy and its temperature falls. A hotter liquid has more particles moving rapidly enough to do this and it will produce more vapour in the space above its surface. Blowing onto hot soup stops these energetic molecules returning, and the soup cools!

If still more energy is put into the liquid, particles will eventually move rapidly enough to produce a vapour with a vapour pressure as large as the pressure above the surface. The vapour is then produced in large bubbles throughout the liquid, and we say that the liquid is boiling. This time the change of state is accompanied by a large increase in the volume of the gas compared with the liquid. (1 cm³ of water will produce 1,800 cm³ of steam.) This means that energy has been supplied to separate the particles against their attractive forces, and we see this as **latent heat** of vaporization.

-ı̇- *Gases, Kinetic theory, Solids*

LOAD

This is the name for a force that a **machine** is acting against. If the force generated by the machine can equal the load and the load is greater than the **effort** then the machine is a *force multiplier*.

-ı̇- *Effort, Machine*

LOGIC GATES

A logic gate is a combination of electronic switches. There are several different types of logic gate, and the output of each type is controlled by the input signals. These signals are **digital**, which means that they can be thought of as either *on* or *off* and are not able to vary between the two extremes. The different gates are called NOT, AND, OR, NAND and NOR. Note that each one has its own symbol.

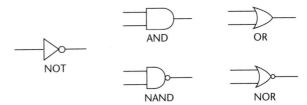

Logic gates

In practice, the gates are contained in **integrated circuits**, which work on 5 V, and the inputs need to be more than 2 V (high) or less than 0.5 V (low) to trigger the gate. Rather than use 'on' or 'high' you will find it easier to use 1; you can then use 0 to mean 'off' or 'low'.

Each gate has a *truth table*, which tells you what will happen at the output for all the possible combinations of inputs. If you look at the name of each gate and compare it with what happens, you should be able to see how each gets its name. Getting that clear in your own mind can be a big help in solving problems. The NOT gate is 'not' what its input is! It is also called an inverter. The AND gate has an output of 1 only when both input A AND input B are 1. The OR gate has an output of 1 when either input A OR input B OR both inputs are 1.

NOT	
Input	Output
0	1
1	0

AND	
Input	Output
0 0	0
0 1	0
1 0	0
1 1	1

OR	
Input	Output
0 0	0
0 1	1
1 0	1
1 1	1

NAND		NOR	
Input	*Output*	*Input*	*Output*
0 0	1	0 0	1
0 1	1	0 1	0
1 0	1	1 0	0
1 1	0	1 1	0

The NAND gate is the same as an AND gate followed by a NOT gate and therefore has all the outputs of the AND gate inverted. In the same way, the NOR gate has all the outputs of an OR gate inverted because it is an OR gate followed by a NOT gate. These last two gates are not examined on all the GCSE syllabuses and you should check your own syllabus carefully to see which are included.

Logic gates can be triggered by the same input sensors that can be used with the **transistor switch** (see **LDR** and **Thermistor**). Outputs can be seen by connecting a **LED** and its resistor between the output and 0 V. The circuits cannot stand a large current, and a **relay** or a transistor driver is needed if the gates are to control another circuit.

Worked example

There is a pattern to solving these, which includes the following steps:

1. Draw a truth table including all the inputs and outputs that are needed.
2. Check if the table belongs to one of the gates that you know. If it does, draw the circuit.
3. If the table does not fit a pattern that you recognize, look for a simple connection between each output and the input columns. If you find one, draw the circuit and check it carefully.
4. If there is no simple connection, look at the inputs that turn on each output and write a 'logic statement'. This will help you to draw a circuit that works but may not find the most simple answer.

Mr Coefield, our French teacher, is very absent-minded and keeps leaving his car headlights turned on when he leaves his car. It would be very helpful if he had a warning light in his car that came on if he opened the door while the lights were on. Design the circuit for him.

Switch A represents the light switch in the car and switch B is fitted in the door so that it comes on when the door is opened. The warning light is an LED on the dashboard. The truth table becomes:

Inputs		Output
		LED
Switch A (lights)	*Switch B (door)*	
0	0	0
0	1	0
1	0	0
1	1	1

You can see that this is the same table as that for the AND gate, so the circuit is a simple one:

Switch A (lights)

Switch B (doors)

LED

Notice that we do not bother to draw in the power supplies to the gate, although they are obviously needed. In a real car you would use a buzzer rather than a LED, it is a better 'attention getter'.

CHECKPOINT

Some animals do not like to be cold or in the dark. Their cage is fitted with two sensors that can light a warning LED. Sensor A reacts to light (1 = dark) and sensor B reacts to temperature (1 = warm). Design a suitable circuit.

✦ LDR, LED, Thermistor, Transistor switch

LONGITUDINAL WAVE

This is a type of **wave** in which the particles vibrate backwards and forwards parallel to the direction of travel of the wave. This causes a series of compressions to travel forward. Sound waves are longitudinal waves.

✦ Wave

LOUDSPEAKER

The loudspeaker has a cone, usually made from card, which can be vibrated. This causes the air in front of it to vibrate, and **sound waves** are produced.

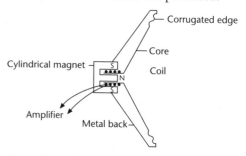

The input is an alternating current of the same frequency as the sound that is to be produced. The current goes through the coil and, since this is between the poles of a magnet, produces an **electric motor effect**. As the current alternates, the coil is vibrated backwards and forwards and the cone vibrates at the same frequency as the current. The edges of the card cone are corrugated so that it can flex many times a second without cracking and splitting.

Strong permanent magnets are necessary to produce loud, clear sounds with little distortion. Small speakers (tweeters) will usually work best at high frequencies and big speakers (woofers) work better at lower frequencies.

✦ Electric motor effect, Sound

MACHINE

It is quite difficult to define a machine, because the word has a wide meaning in everyday use. In physics, a machine is a device that does some **work**. While doing this it will also transform some **energy**.

There are many different machines, **levers** and pulleys being simple examples. There are a number of ways of trying to decide whether a machine is good at its job. One of them is to calculate its **efficiency**, but if speed is important you might be more interested in **power**.

The force that is put into the machine is called the *effort* and the force that is overcome by the machine is called the *load*. We can get a measure of how useful the machine is by working out

$$\text{mechanical advantage} = \frac{\text{load}}{\text{effort}}$$

The *mechanical advantage* (M.A. for short) has no units. It is just a number that tells you how many times bigger the load is than the effort. Many machines are designed to be *force multipliers*, and the M.A. tells you how many times the effort will be multiplied. A car jack will have a big M.A., because it has to lift a load that is much larger than the effort that you are putting in. The M.A. is usually less than you might expect, because the effort has to overcome other forces as well as the useful one. There is often friction in the machine, for example, and some of the effort and energy put in is used up in overcoming that. If you can use oil to lessen the friction, the effort needed is smaller and the M.A. is increased.

> *Remember: Few machines are anywhere near 100 per cent efficient (remember friction etc.), and this can be a useful check on your answer when solving problems.*

⊹ *Efficiency, Force, Lever*

MAGNETIC FIELDS

A magnetic field is the **force** that is exerted on a magnetic material in the space surrounding a magnet. The strength of the field gets weaker as the distance from the magnet increases.

We can show a picture of a magnetic field by drawing 'lines of force'. Each line is the path that would be taken by a free N pole. (Since you cannot have a N pole without a S pole it is impossible to have a truly free N pole, but the idea is useful because it helps you to 'see' the field.) Each line of force will be going away from N poles and towards S poles. The lines cannot cross, because the N pole can be moved by the field in only one direction.

There are several ways to plot these lines. The quickest and easiest way is to use iron filings sprinkled evenly on to a piece of paper placed over the part of the field that you wish to examine. Each iron filing becomes a tiny magnet and will turn to line up with the field if you gently tap the paper. The method is not very good in strong fields, because the iron filings get pulled across the paper towards the nearest strong pole!

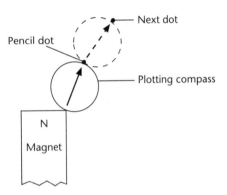

A better method is to use a 'plotting compass', which is a small magnetic compass. Put a piece of plain white paper in the required part of the field, place the compass on the paper and put a pencil dot on the paper as close as possible to the N of the compass. Move the compass so that the S of the compass is next to the dot and draw a second dot near to the N end. Continue this process as far as needed and then join the dots. Do this, beginning in different places, until you have enough lines to show the field.

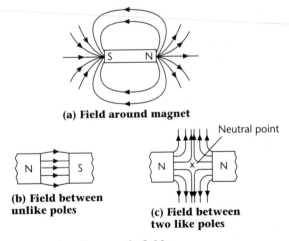

(a) Field around magnet

(b) Field between unlike poles

(c) Field between two like poles

Some examples of magnetic fields

> *Remember: The lines are closer together where the field is stronger. At a neutral point all the magnetic forces 'cancel' – in the case shown there are two equal and opposite forces.*

-**+**- *Magnetic materials, Magnetic poles and magnetic forces*

MAGNETIC MATERIALS

Only iron, cobalt and nickel and the alloys that contain them can be magnetized. This means that they can exert a **force** on other magnetic materials that are near to them. The force does not need a material to exist in, and magnets work perfectly well in a vacuum. The force decreases as the distance from the magnet increases, and we say there is a **magnetic field** around the magnet that decreases with distance.

Hard magnetic materials are difficult to magnetize but, once magnetized, will also be difficult to demagnetize. These materials are useful for permanent magnets, and a lot of research goes into producing strong permanent magnets for such things as **loudspeakers** and **electric motors.** Steel is a hard magnetic material, as are some special alloys such as 'alnico'.

Soft magnetic materials are easy to magnetize but also easy to demagnetize. These materials would be used for the cores of electromagnets and the cores of **transformers**. Iron is a fairly soft magnetic material.

-**+**- *Electric motor effect, Electromagnetic induction, Loudspeaker, Magnetic poles and magnetic forces, Transformer*

MAGNETIC POLES AND MAGNETIC FORCES

On each magnet there will be two areas where the magnetic force is at its strongest. These places are often at opposite ends or sides of the material and they are called *poles*. If the magnet is suspended on a thread so that it can turn freely it will turn until the poles are approximately north–south. The poles are then called a *north-seeking pole*, which we shorten to north pole or N, and a *south-seeking pole*, shortened to south pole or S. There will always be one pole of each type, you cannot have a single magnetic pole. If we suspend two magnets close to each other we find that they obey two simple rules:

● like poles repel

● unlike poles attract

A small compass can also be used to identify the poles on a magnet. Make sure that you know which is the N pole of the compass needle and bring it close to each end of the magnet and watch to see which pole *repels* it. This is the N pole of the magnet. Be careful, because a compass needle will be attracted to both ends of an unmagnetized piece of magnetic material, so you must look for repulsion and *not* attraction.

A magnetic field is produced by an electric **current** in a wire, and sending through a long coil with a lot

of turns can produce a strong magnetic effect. This sort of coil is called a *solenoid*. There are two types of use for this:

1. Put a soft iron core in the coil and it becomes magnetized, making the magnetic field from the coil much stronger and producing an *electromagnet*. You can sort out which pole is which if you follow the rule illustrated in the figure. Look into the ends of the coil and imagine the current flowing from + to −. The direction of the current fits one of the N or S symbols as shown. An electromagnet like this is used in the **relay** and in large electromagnets for lifting scrap metal. In some cases, the direction of the field does not matter and you can use an **alternating current** (a.c.) instead of **direct current** (d.c.). An electromagnet will still attract scrap metal, even though its poles are 'swopping over' rapidly!

Solenoid

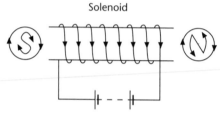

2. If you set up a solenoid as in the diagram and put a piece of magnetic material inside it, the material becomes magnetized. It works better if you put a variable resistor in the circuit, slowly increase the current (and the field) up to a maximum and then decrease back to zero rather than suddenly switching the current on and off. This is the best way to make a magnet. If you have not got a coil and a power supply you can weakly magnetize a material such as a needle by stroking it repeatedly in one direction with a permanent magnet.

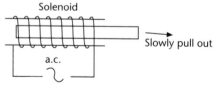

(a) Demagnetizing rod

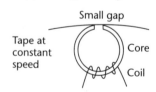

(b) Tape recorder head

3. If you use the same circuit with an a.c. supply the material inside the coil is demagnetized. This works better if you slowly withdraw the material from the coil to a distance with the a.c. turned on. Many delicate instruments – even mechanical watches! – need to be demagnetized in this way so that they work accurately.

Tape recorder heads are a good illustration of

magnetism in action. The recording head is an electromagnet that carries a current whose **frequency** and strength follows the sound to be recorded. The tape is a thin plastic backing tape with a magnetic material on one side. As the tape passes the head, a magnetic pattern is recorded in this magnetic layer. It can be erased by sending it past the erase head, which is another electromagnet. The erase head carries a high-frequency current, so its rapidly changing field demagnetizes the tape. The playback head can reproduce the sound when the tape goes past it because a changing magnetic field in a coil induces a current in it (see **Electromagnetic induction**). Information is recorded on magnetic disks for use in computers, and the process is similar to tape recording, except that the information is stored in circles on the disk called tracks.

-+- **Electric motor effect, Electromagnetic induction, Relay**

MANOMETER

The U-tube manometer is a way of measuring the **pressure** of gases.

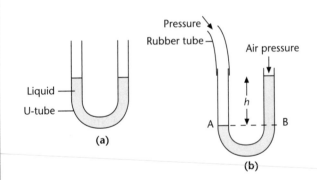

If you think about the tube in the diagram, you can see that the same air pressure acts down on the liquid in each arm of the tube. The liquid will balance at the same level in each arm. If you now connect a gas pressure to one side, a larger pressure will be pushing down on one side. The pressure at the level of the lower surface must be the same on both sides, or the liquid would be pushed to one side or the other. In the diagram, the pressure on the surface A is the same as the pressure at the same level B.

$$\text{pressure at A} = \frac{\text{pressure of}}{h\,\text{mm of liquid}} + \frac{\text{atmospheric}}{\text{pressure}}$$

If you do this with water in the tube and connect it to the gas supply it will show you that the gas pressure is about 220 mm of water pressure above atmospheric.

If you measure pressures lower than atmospheric the liquid moves the other way. For larger pressures, the tube can be filled with a denser liquid.

> *Remember: (1) It is VERTICAL height that causes the pressure in a liquid. If you tip the manometer sideways the liquid may well get nearer to the end of the tubes but the vertical heights will remain the same. (2) The width of the tubes does not matter and can even be different on each side. A wider tube will contain a greater WEIGHT of liquid, but the FORCE is spread over a larger area and the pressure remains the same!*

-+- **Density, Pressure**

MASS

Mass is a measure of the quantity of matter in a material. It is measured in kilograms; a block of a platinum–iridium alloy kept at Sèvres in France is the standard kilogram for SI units. It is important to note that mass is *not* the same as **weight** and is *not* a **force**. The total mass of an object will remain the same wherever you take it. Even when a space probe is a very large distance from any star or planet and becomes weightless it will still have the same mass.

Mass is also a measure of the *inertia* of a body. Inertia is a measure of the resistance of an object to having its motion changed. A larger mass will need a larger force to produce the same **acceleration** because it has more inertia. **Newton's second law** shows the connection between mass and forces.

We sometimes use an additional unit, the tonne, for large objects.

$$1 \text{ tonne} = 1{,}000 \text{ kg}$$

If you find it difficult to imagine a 1 kg mass remember that it is the mass of a bag of sugar from the supermarket!

-+- **Appendix One, Density, Weight**

METRE

-+- **Length**

MICROWAVES

-+- **Electromagnetic spectrum**

MILKY WAY

The **Sun** belongs to the Milky Way, which is a spiral galaxy, about 100,000 light years across, containing about 100 billion stars.

-+- **Galaxy**

MIXTURE

Mixtures are substances that contain various amounts of **elements** and/or **compounds** mixed together. They can be physically separated easily. For example, a mixture of iron and sulphur can be separated easily with a magnet.

MOLECULE

All matter is made up of *particles*; there are three different types: **atom**, molecule and **ion**. A molecule is a particle that contains two or more atoms chemically joined together. Molecules can contain the same type of atom, or different atoms chemically joined (bonded). Each molecule has a name and a chemical formula to represent which atoms are joined together. For example:

Molecule	Name	Formula
H — H	hydrogen	H_2
H \O H/	water	H_2O

MOMENT

The moment of a **force** is a measure of its turning effect. The force will act on an object that rotates about a *pivot* (or *fulcrum* or *axle*) and the size of the moment can be found from

moment = force × perpendicular distance to pivot

The perpendicular distance is used because it is the shortest. The units are Nm, assuming that you use N for the force unit and m for the distance.

> Remember: The units are not N/m.

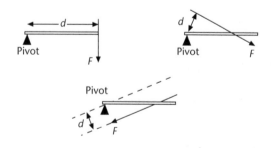

Moment = Fd

Imagine that the pivot in the figure is the hinge of a door. You will get the largest moment if you push at 90° to the door and as far from the hinge as possible. A door can be quite difficult to close if you push close to the hinge or at a small angle to the surface, because the moment is small.

Many objects have turning forces acting on them but do not rotate, because the moments of all the forces balance. An object on which all the forces balance is in **equilibrium**. This is stated in the *Principle of Moments*:

> If an object is in equilibrium, the sum of the clockwise moments is the same as the anticlockwise moments. This will be true if you calculate the moments about any point on the object.

It is usually easiest to calculate the moments about the pivot. Find each of the clockwise moments first and then add them. Do the same with the anticlockwise moments and then put the two totals equal to each other. Any bit of missing information can then be found.

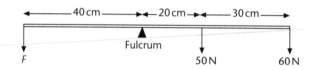

Find the force F that will keep the beam in the above diagram balanced. The beam is balanced at its **centre of mass** so you can ignore its weight.

Taking moments about the pivot:

$$\text{clockwise moments} = 50 \times 0.2 + 60 \times 0.5$$
$$= 10 + 30$$
$$= 40\,\text{Nm}$$
$$\text{anticlockwise moments} = F \times 0.4\,\text{Nm}$$

Using the principle of moments:

$$\text{anticlockwise moments} = \text{clockwise moments}$$
$$F \times 0.4 = 40$$
$$F = \frac{40}{0.4} = 100\,\text{N}$$

You can easily check this principle using a 1 m rule with weights hanging from it. You need to balance it at its mid-point so that its weight does not matter (the distance from the pivot to the weight is then 0 m, so the moment is also 0!). You can do this by balancing the rule at the 50 cm mark or by drilling a hole through it at its centre and balancing the rule on a pin through the hole (see a teacher *before* you drill the hole!). Remember that your measurements must be from the pivot and *not* just the markings on the scale.

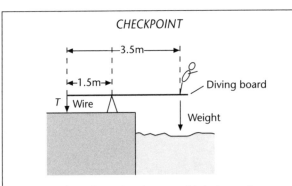

CHECKPOINT

A diving board is 3.5 m long and is balanced as shown in the diagram. It is stopped from tipping over by a thick wire at one end. What extra tension **T** in the wire will be caused by a swimmer of weight 600 N standing at the other end of the board?

-+- *Force, Lever*

MOMENTUM

The momentum of a body can be found from

$$momentum = mass \times velocity$$
$$M = mv$$

The units for momentum are kgm/s.

For example, what is the momentum of a 20 tonne lorry moving at 15 m/s?

$$momentum = mass \times velocity$$
$$= 20,000 \times 15$$
$$= 300,000 \, kgm/s$$

Newton's second law shows the connection between **force** and the change of momentum.

> Remember: A small mass moving very quickly can have a very large momentum. Collisions with bullets can show that the bullet has a lot of momentum and kinetic energy.

-+- *Conservation of momentum, Force, Mass, Newton's laws*

MOON

The moon is a **satellite** that takes about 28 days (a lunar month) to orbit the Earth. The distance between the Earth and the Moon is approximately 400,000 km, and it is held in its orbit by gravitational attraction between it and the Earth. The Moon also rotates on its own axis every 28 days, so the same side of the Moon faces the Earth all the time. It has no atmosphere and no water.

MOTION

-+- *Equations of motion*

NAND GATE

-+- Logic gates

NATIONAL GRID SYSTEM

One reasonably cheap way of distributing **energy** all over the country is to use the fuel to produce electricity. The electricity is then put into the National Grid System so it can be used wherever it is needed.

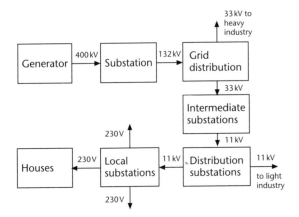

Electricity distribution

The main problem is that electric currents waste energy by heating the cables. The power wasted is given by energy = I^2R and the power transmitted is equal to VI (see **Heating effect of electric current** and **Power**). We can therefore waste less heat by transmitting the power at high voltage and low current. This will need to be changed back to a lower voltage and higher current at the 'receiving' end of the cable. The changes in voltage are done in stages. The power is generated at 25 kV, and this is increased to 400 kV for the cables linked across the country. This is changed down to 132 kV between towns and then down to 33 kV at a grid distribution point, where it is sent out to the local area and to some industries that have their own substation. It is then lowered again in intermediate substations to 11 kV and finally to 230 V in local substations. The voltage has been kept as high as possible throughout this process to reduce the heat losses as far as possible. Somewhere near your home you will find the local substation, usually painted green. Take care and *look* only from the outside – the voltages here are very dangerous.

To make these changes in voltage we use a **transformer**. A step-up transformer increases the voltage at the transmitting end of the cable and a step-down transformer reduces the voltage at the

receiving end. Transformers do *not* work on d.c., so our supply current must be a.c.

> Remember: Transformers are quite efficient and heat losses are reduced but a lot of energy is always wasted in changing from the chemical energy of the fuel to the electrical energy – usually in waste heat.

Worked example
Find the power wasted in transmitting 5 kW along a 4 Ω cable at (1) 250 V and (2) 25,000 V.

1. current = $\dfrac{\text{power}}{\text{volts}}$ From power = VI

$$= \frac{5,000}{250} = 20 \text{ A}$$

power wasted in cable = $I^2R = 20^2 \times 4$
$$= 1,600 \text{ W} = 1.6 \text{ kW}$$

2. current = $\dfrac{\text{power}}{\text{volts}}$

$$= \frac{5,000}{25,000} = 0.2 \text{ A}$$

power wasted in cable = $I^2R = 0.2^2 \times 4 = 0.16 \text{ W}$

The example is exaggerated, but look at the waste in (1) – it would be far too inefficient – and compare it with (2), where the waste is so small that you probably would not be able to measure it.

-+- **Heating effect of electric current, Power, Transformer**

NEUTRON

A neutron is one of the three particles from which **atoms** are made. It has no **charge** (neutral) and a **mass** of one unit. It is found in the nucleus of atoms.

-+- **Atom, Electron, Proton**

NEWTON

This is the unit of measurement for a **force**: symbol N. A force of 1 newton gives an **acceleration** of 1 m/s^2 to a mass of 1 kg.

> Remember: To get an idea of the size of 1 newton, picture a bag of sugar. This will usually have a mass of 1 kg and will weigh 10 N on Earth.

-+- **Newton's second law**

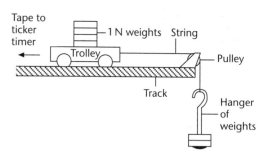

NEWTON METER

-✧- *Hooke's law*

NEWTON'S FIRST LAW

The law states that a body will stay at rest, or continue at uniform **velocity**, unless it is acted on by an external **force**.

Although the law does not look very useful at first, it tells you what happens to an object that has *no* force acting on it, so you can compare it with what happens when a force *is* acting on it. It means that an object far out in space will continue on in a straight line at the same speed for ever unless a force acts on it to change its velocity. This implies that forces will change velocity – i.e. cause **acceleration** – and leads to the idea of **Newton's second law.**

On Earth, all objects have at least one force acting on them – weight – and it must have been very difficult for people such as Galileo and Newton to discover what now looks to be a simple law!

-✧- *Force, Newton's second and third laws*

NEWTON'S SECOND LAW

The **force** applied to a body is directly proportional to the rate of change of **momentum** in the direction of the force. So the law is:

$$\text{force} \propto \frac{\text{change of momentum}}{\text{time taken}}$$

and we can show that this is the same as

$$\text{force} \propto \text{mass} \times \text{acceleration (in the direction of the force)}$$

This makes it possible to define a unit for force by putting **mass** = 1 and **acceleration** = 1 and making force = 1:

One *newton* is the force that will accelerate 1 kg of mass at 1 m/s² in the direction of the force.

If we use this new unit of force, kg for mass, m/s for velocity and m/s² for acceleration, the law becomes equations:

$$\text{force} = \text{mass} \times \text{acceleration} \qquad F = ma$$

$$\text{force} = \frac{\text{change of momentum}}{\text{time taken}} \qquad F = \frac{M_1 - M_2}{t}$$

You can show that force is propotional to acceleration using a **ticker timer** and a trolley. Set up the apparatus as in the figure. You should put small wedges under the start of the track unit until the trolley will run down at a constant speed after a small push. This can be checked by using the ticker timer, which should then produce equally spaced dots. Put about six 1 N weights (100 g masses!) on the trolley. Hang one of the weights on the string as

shown in the diagram, start the timer and release the trolley. From the dots on the tape, work out the acceleration and make a note of this and the accelerating force in a table of results.

Transfer each of the masses in turn from the trolley to the hanger and repeat the experiment each time, recording the results in your table. Notice that by transferring the weights from the trolley you keep the total mass constant and only the force and acceleration change. It is a good idea to repeat all the results now by taking the weights off one at a time to check all your answers.

You can now plot a graph of acceleration against time. This should be a straight line as shown in the diagram and proves that the force is proportional to the acceleration. To complete the law you would need to do a second experiment in which you keep the accelerating force constant and change the total

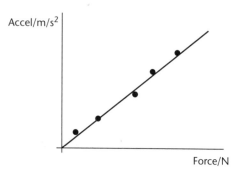

mass of the trolley and weights each time. The results show that mass × acceleration is a constant.

The same experiment can also be carried out using light sensors and a computer instead of the ticker timer. A card attached to the trolley will go past two light sensors. The speed at each sensor and the time taken to go from one to the other are measured, the computer program then works out the acceleration. The graph of acceleration against time will be straight, as in the ticker timer experiment.

Worked example
A car accelerates at 4 m/s² and has a mass of 1,000 kg. What is the average force being produced by the engine?

$$\begin{aligned}\text{force} &= \text{mass} \times \text{acceleration}\\&= 1,000 \times 4\\&= 4,000 \text{ N}\end{aligned}$$

CHECKPOINT

(1) If a 2,000 N force acts on a 100 kg spacecraft, what acceleration can it produce?
(2) If the force lasts for 10 s, what is the change in velocity?

A lorry of mass 25,000 kg is moving at 20 m/s. It brakes to a stop in 20 seconds.

(3) What is the acceleration of the lorry?
(4) What is the braking force acting on the lorry?

You will find different examples in the entries on *force* and *Newton's third law.*

-✦- *Force, Momentum, Newton's first and third laws*

NEWTON'S THIRD LAW

If two objects act on each other (e.g. collide or are thrown apart), both experience the same size of *force* in the same straight line but in opposite directions.

Examples

1. When a gun is fired, the bullet has a force sending it forwards and a force the same size sends the gun backwards. We call this the 'recoil' and it causes great problems with a big gun. Modern artillery is carefully designed to absorb the recoil, but it is thought that some early wooden battleships turned over when all the cannon were fired together!
2. If water is squirted from a hosepipe there is a reaction force pushing the pipe backwards. This can be a problem with large fire hoses.
3. A *rocket* carries with it all the fuel and oxygen that it needs. When this is burned the hot expanding gases can only leave at the back of the engine. This is the same situation as the hosepipe – the rocket is driven forward by the reaction force.
4. A *jet engine* works on exactly the same principle as the rocket, because hot (very fast) gases leave the rear of the engine and cause the engine to be forced in the opposite direction. It is designed to work only in the Earth's atmosphere, and it does not need to carry a supply of oxygen with it. The air is taken in at the front of the engine and compressed before being burned with the fuel. The compressor can be driven by a turbine in the path of the hot gases before they leave the engine.
   Note that rockets and jet engines (and hosepipes!) will produce this force in obeying the third law. They do *not* need something to push against – a rocket, for example, will work just as well in a vacuum.
5. Stepping from a small boat or a canoe to the bank can give an interesting example of the third law, as the force moving the boat away from the bank is the same as the force moving you towards the bank!

This reaction force from the third law can usually be worked out from the equations that we found from the second law. (Find the force on the cannon ball and reverse the direction for the force on the cannon.) It is sometimes also useful to know that:

$$\text{force} = \frac{\text{change of momentum}}{\text{time taken}} = \frac{\text{mass} \times \text{velocity}}{\text{time taken}}$$

$$= \frac{\text{mass}}{\text{time taken}} \times \text{velocity}$$

So you can work out the force on a jet, rocket or hosepipe by finding the mass per second that is leaving the back and its velocity and then multiplying the two.

Worked example
A hosepipe squirts water at 2 kg/s and a velocity of 10 m/s. What is the force on the pipe?

force = mass/second × velocity
= 2 × 10
= 20 N

-✦- *Force, momentum, Newton's first and second laws*

NOISE

Noise is unwanted sound. It may be musical (and too loud!) but is often an irregular or random mixture of sound.

The loudness of noise is measured in decibels (dB). Zero dB is just about audible (a sound pressure of 2×10^{-5} Pa), and each increase of 20 dB means that the pressure of the sound wave has increased by a factor of ten. Our ears do not react linearly to this pressure, and it is easier to remember that each increase of 10 dB doubles the loudness. Some examples may help:

- 30 dB   a whisper
- 60 dB   normal conversation
- 80 dB   door slam
- 95 dB   noisy factory

A good car silencer can reduce exhaust noise by 60 dB.

-✦- *Appendix One*

NOR GATE

-✦- *Logic gates*

NOT GATE

-✦- *Logic gates*

NUCLEAR ENERGY

Nuclear energy is the **energy** that binds together the **nucleus** of an atom. If the nucleus of an atom is split into two parts, the two nuclei may need less binding energy than the original big nucleus, and the extra energy is released as heat. We see this as a *very* tiny quantity of nuclear mass turning into energy.

-⧈- *Energy, Nuclear reactor*

NUCLEAR REACTOR

A nuclear reactor produces energy from a controlled **chain reaction**. The reaction involves the **fission** of atoms of uranium-235 or plutonium-239. As each nucleus is split into two parts a small fraction of the mass is turned into energy. This energy appears as heat, which is steadily removed from the reactor core and used to boil water to make steam. The steam is then used in the same way as in other types of power station (i.e. it drives turbines, which in turn drive generators to produce the electricity). The main parts of the advanced gas-cooled reactor (A.G.R.) are shown in the figure. They have the following purposes:

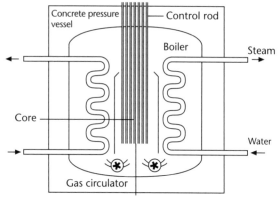

Advanced gas-cooled reactor

Fuel rods

The fuel rods or pins are made from stainless steel and contain the uranium or plutonium fuel. They are lowered into the centre of the reactor (the reactor core) and can be taken out again when the fuel in them is used up ('spent'). The rods will then contain the 'fission products', which include the new nuclei produced when the fuel atoms split. These new isotopes will include many dangerously radioactive ones and the spent fuel needs careful treatment. The rods are first put into a large pool of water and left there until the short half life nuclides have decayed. Water is used because it is quite good at absorbing the neutrons that are being produced. The rods are then stripped open by remote control in a sealed workshop and the remaining chemicals dissolved in

acid so that they can be chemically separated. The long half life isotopes that are still dangerous are then stored until they can be safely disposed of. One way of doing this is to seal the radioactive material in glass or ceramic bricks so that it cannot leak away and then bury the bricks in caves in a stable rock.

Moderator

The neutrons that are produced in the fission have too much kinetic energy for them to be easily absorbed and produce another fission. The moderator, which is usually graphite or water, slows down the neutrons so that they are more likely to be captured by a ^{235}U nucleus. The slower neutrons are called thermal neutrons, and a reactor that operates using these slower neutrons is a *thermal reactor*.

Control rods

The chain reaction has to be controlled so that the reaction is just 'critical', i.e. it is just keeping itself going. The control rods are made from cadmium or boron steel, which are good at absorbing neutrons. If the reaction goes too fast, the rods are lowered further into the reactor so that they absorb neutrons and the reaction is slowed down. The reactor will have a lot of this type of rod, so that they can also be put into the core and shut the reactor down in any sort of emergency.

Coolant

The purpose of the reactor is to produce heat, and this heat must be steadily removed from the core as it is produced by the chain reaction. The heat is removed by sending a coolant through the core and then sending the coolant through a heat exchanger. The coolant may be carbon dioxide or water (under pressure so that it does not boil), and it is circulated in a closed circuit so that radioactivity is not transferred to the outside.

Natural uranium contains only about 0.7 per cent uranium-235, the rest being uranium-238, which does not fission. The uranium fuel has to be processed to increase the proportion of ^{235}U and this is expensive, since both isotopes will have identical chemistry. The uranium-238 can be put in a 'blanket' round the core of a *fast breeder* reactor. It will then absorb some fast neutrons, emit β radiation and become plutonium-239, which can be used as a fuel. Some plutonium can also be extracted from spent fuel rods. Plutonium reactors are smaller and more efficient, but plutonium is *very* toxic.
    The materials of the reactor will absorb neutrons and become radioactive. (Materials can deliberately be made radioactive in this way – isotopes for medical use, tracers, genetic research, etc.) The reactor must be shielded to prevent radiation leaving

and should be contained in a pressure vessel that can withstand accidents and prevent the loss of radioactive material. The coolant circuit should be completely closed for the same reason. The proportion of fissionable material in the core of a reactor is not large enough to cause an explosion after an accident, but it can produce enough heat to cause a 'meltdown'. The pressure vessel containing the core should be able to withstand this for a while but, because of the huge quantities of energy available, there would be serious consequences. Modern reactors are designed to be shut down very quickly in the case of accidents such as a loss of coolant.

A properly managed reactor can produce a lot of electricity at a competitive price with little or no pollution to the air or surrounding countryside. It does not require very large quantities of chemical fuel and does not produce large waste-tips or spoil heaps. The waste products are the main problem in that they have long half lives and present long-term problems in safe storage or disposal.

The other main problem is the expensive technology involved in dismantling the reactor after its useful life.

-⟡- *Energy, Fission*

NUCLEON NUMBER

The nucleon number is the total number of **protons** and **neutrons** in the nucleus of an **atom**.

-⟡- *Atom, Isotope*

NUCLEUS

A nucleus is the central part of an atom and is made up of **protons** and **neutrons**.

-⟡- *Atom, Nucleon number*

NUCLIDE

A nuclide consists of atoms in one particular **isotope** of an element.

-⟡- *Atom, Isotope*

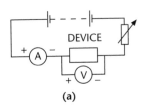
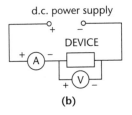

(a)   (b)

OHM

The ohm is the unit for **resistance**.

A resistor has a resistance of 1 ohm if a potential difference of 1 V across it will drive a current of 1 A through it. The symbol for ohm is Ω.

✛ *Ohm's law, Resistance*

OHM'S LAW

If a **resistor** is kept at a constant temperature, the **current** flowing through it is directly proportional to the **potential difference** between its ends.

If we use the **ampere** as our unit for current, the **volt** as our unit for **potential difference** (p.d.) and the **ohm** as our unit for **resistance**, the law gives us the connection between them:

$$\text{resistance} = \frac{\text{potential difference}}{\text{current}}$$

$$R = \frac{V}{I} \quad \text{or} \quad V = IR$$

Worked example

What is the resistance of a resistor that takes a 20 mA current from a 9 V battery?

$$\text{resistance} = \frac{\text{potential difference}}{\text{current}}$$

$$= \frac{9}{0.02} = 450\,\Omega$$

CHECKPOINT

What current will flow through a bulb of resistance 6 Ω when it is connected to a 12 V supply?

You should notice that Ohm's law will be fully obeyed only by metallic conductors that remain at a constant temperature. If the temperature rises, the resistance will also increase, as the current has to find its way between metal **ions** that are vibrating faster. This is why **fuses** are more likely to blow at the instant a circuit is switched on – the components are cooler and allow slightly more current through until they warm up. Some devices do not obey Ohm's law, and you can use the circuit illustrated to find out what happens in a particular device.

Either of the circuits in the figure will do. You may need to find instruments with a suitable scale – e.g. a milliammeter instead of the **ammeter** if the device has a large resistance. Note how the meters are connected. You can get a quick check of resistance by measuring p.d. and current and dividing, but this is often inadequate. Take a set of readings for V and I by changing the variable resistor in (a) or the voltage from the power supply in (b). Allow time after each change for the device to reach a constant temperature. Plot a graph of V on the Y-axis against I on the X-axis. You can reverse the device to get some readings in the negative direction if you wish – this may make a difference for some devices!

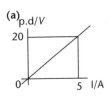

Resistance = gradient   **Increasing resistance**   **Sudden resistance**
= 20/5 = 4 Ω   **in a bulb**   **change in a diode**

Decreasing resistance in a thermistor

A straight line shows a device such as a **resistor** or a coil of wire obeying Ohm's law. The resistance can be found from the gradient of the graph as shown. This is much better than one value, as it gives an average of several results and clearly shows up any **rogue results**.

A line that is not straight shows that the device does not obey Ohm's law. A graph of the results from a 12 V bulb curves upwards, showing the increase in resistance as the filament of the bulb gets hotter.

The **diode** in graph (c) does not obey Ohm's law, because it does not conduct when it is reverse-biased. The **thermistor** in graph (d) will have a lower resistance when it is warmed by the current. There is the basis for an interesting investigation here if you put the thermistor into a beaker of water and find out how its resistance changes with the temperature of the water – you might also use the results to design a thermometer.

✛ *Current, Potential difference, Resistance*

Op Amp

Operational amplifiers are **integrated circuits** that have a number of uses. Two inputs control a single output, one of which is the inverting input, marked −, the other the non-inverting input, marked +. The device needs *both* a + and a − supply (about ±12 V) and the centre of these is taken as 0 V when measuring inputs and outputs.

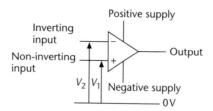

If the inverting input in the figure is connected to 0 V and a small positive voltage V_1 applied to the non-inverting input, an amplified positive voltage is produced at the output. If the non-inverting input is connected to 0 V and a small positive voltage V_2 applied to the inverting input, the output voltage is still amplified but is negative, i.e. it has been inverted. The amplification of the device if it is used in this way is very large:

$$\text{voltage gain } (A) = \frac{\text{voltage output}}{\text{voltage input}} = \text{approx. } 100,000$$

Note that the value of A is a number without units.

A much better amplifier can be made by using *negative feedback*. This uses the inverting input with the non-inverting input connected to 'ground' (0 V).

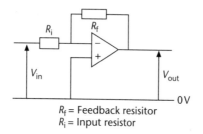

R_f = Feedback resisitor
R_i = Input resistor

Some of the inverted output is put back into the input – this is the 'feedback' and is negative because it has been inverted when compared with the input voltage. This makes the voltage gain much smaller, but it now has a known value and is constant over a much wider range of frequencies.

In the circuit in the second diagram;

$$A = \frac{V_{out}}{V_{in}} = \frac{-R_{feedback}}{R_{input}}$$

This makes the gain independent of the particular device used, but remember that the maximum V_{out} cannot exceed the supply voltage and will be inverted.

Other uses include the *voltage follower*, in which the output copies the input but is 'buffered' from it so that the device connected to the output is protected against large voltage inputs; and the *voltage comparator*, which produces an output of either ± supply voltage depending on the input.

Worked example

A voltage amplifier is to be built from an op amp using negative feedback. An output voltage of –10 V is to be produced from an input of 3 V. The input resistor has a value of 45 kΩ.

1. Draw a diagram of a suitable circuit.
2. What is the voltage gain?
3. What will the value of the feedback resistor be?

1. See previous figure.

2. $A = \dfrac{V_{out}}{V_{in}} = \dfrac{-10}{3}$

   $= -3.33$

3. $A = \dfrac{-R_{feedback}}{R_{input}}$

   $3.33 = \dfrac{R_{feedback}}{45,000}$

$R_{feedback} = 3.33 \times 45,000 = 150,000\,\Omega = 150\,k\Omega$

✚ *Integrated circuit*

Optical Centre

✚ *Lens*

Or Gate

✚ *Logic gates*

Oscilloscope

An oscilloscope is an instrument that is used in electronics to measure **voltages** and to show how they change with time.

The instrument has a bright spot focused on its screen that is caused by a beam of electrons hitting a layer of phosphor on the inside of the screen. This beam travels between two sets of deflectors that can move it up and down or left and right, depending on the direction of the voltage applied. Larger voltages move the spot a proportionally greater distance. So that you can use the instrument with a wide range of voltages, the input voltages go through an amplifier. This amplifier has a rotary switch that you can set to the number of volts that you want for each cm of movement on the screen. To help you measure distances, the screen is usually marked with a grid rather like a piece of graph paper.

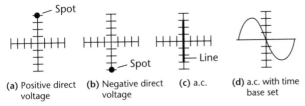

(a) Positive direct voltage **(b)** Negative direct voltage **(c)** a.c. **(d)** a.c. with time base set

Oscilloscope traces

Measuring p.d. in a d.c. circuit

Adjust the spot to the centre of the screen using X and and Y shift controls. Make sure that it is correctly focused and reasonably bright (using the focus and brightness controls). Switch the Y-amp control to a suitable number of V/cm. Connect the **potential difference** (p.d.) to be measured to the Y terminals. The spot will move up and stay at its new height. Measure how far it goes and multiply by the number of V/cm.

Worked example

When a direct p.d. is applied to the Y terminals of an oscilloscope the spot moves up by 4.5 cm. If the Y-amp is set at 2 V/cm what is the applied voltage?

$$\text{applied voltage} = \text{V/cm} \times \text{distance moved}$$
$$= 2 \times 4.5$$
$$= 9 \text{ V}$$

Measuring p.d. in an a.c. circuit

Adjust all the controls and connect up as in the previous section. Since the spot is being driven repeatedly up and then down the screen, it will bounce up and down in the centre of the screen. At 'mains' frequency (50 Hz) the spot is going up and down so rapidly that a vertical line is drawn on the screen. From the *centre* to the top or bottom of the line represents the 'peak voltage'.

Worked example

An alternating p.d. connected to the Y inputs of an oscilloscope produced a line 8 cm from top to bottom with the Y-amp. Set at 5 V/cm. What is the peak voltage?

$$\text{peak voltage} = \text{V/cm} \times \text{distance moved}$$
$$= 5 \times 4 = 20 \text{ V}$$

> *Remember: The main advantage of the oscilloscope when measuring p.d. is its high input impedance. This means that it will take almost no current at all from the circuit to which the Y inputs are connected and therefore the circuit values are not changed.*

Displaying a waveform

The voltage to be displayed is connected to the Y-terminals as described in the sections above, making sure that the V/cm control is on a suitable setting. The spot can now be made to move steadily across the screen in a fixed time and then 'fly back' in a very short time and repeat this movement continuously. This is done by turning on the *time base*. There will be a time base or velocity control that adjusts the speed of the spot across the screen, and this can now be changed until the waveform appears clear and stationary on the screen. If you have set the time across the screen to be the same as that for the spot to make one wave, you will see one complete wave. If the time is set to equal twice the 'periodic time' of the wave, you will see two complete waves, and so on. Many oscilloscopes have the time base control calibrated so that you read off the frequency of the wave that you are watching.

✛ *Alternating current, Potential difference*

PARALLEL

Components of a circuit will be in parallel if the **current** in the circuit divides, part going through each component, and then joins up again later to complete the circuit. This means that there is more than one possible route for the current to flow betwneen two points in the circuit.

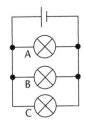 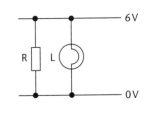

In the figure the bulbs A ,B , and C are all in parallel. This circuit is useful because if one bulb fails the other bulbs can remain on. The **resistor** R and the lamp L are also in parallel. Putting identical cells in parallel produces a bigger **cell** of the same voltage, which has the ability to provide larger currents than one cell could.

-∔- **Current, Resistor, Series**

PASCAL

In measuring **pressure**, 1 N/m^2 is called 1 pascal (Pa).

PAYMENT FOR ELECTRICAL ENERGY

The unit for **energy** is the **joule**, and this is the unit used in the equation for the **heating effect of electric current**. It takes 4,200 joules to raise the temperature of 1 kg of water by 1 °C, so the number of joules to heat the water for a bath is very large. Your electricity bill would have some huge numbers at a very tiny cost for each one by the end of each quarter! We can make things much easier by inventing a larger unit. Instead of the joule, which is 1 watt for 1 s, we can use the *kilowatt-hour*, which is 1 kW for 1 hour. If you work it out, you will find that 1 kWh = 3,600,000 J = 3.6 MJ. This is a better size − 1 'unit' costs 6 p at the moment.

To find the cost of using electricity, work out the number of units by multiplying the **power** in kW by the time in hours and then multiply the answer by the cost per unit.

Worked example
Mr Speed left the computer system at work switched on by mistake. The system is rated at 400 W and was left on for 10 hours. What did this cost at 6p per unit?

number of units = 0.4 × 10 = 4
cost = units × cost per unit = 4 × 6 = 24p

You can add together the power ratings for appliances that are all on for the same time and find the cost in one calculation.

CHECKPOINT

A school uses 120 lights for an average of 6 hours each day. If each light is rated at 100 W, and a unit of electricity costs 6p, what does this cost for a week?

-∔- **Heating effect of electric current**

PERIODIC TIME

-∔- **Frequency**

PLUG

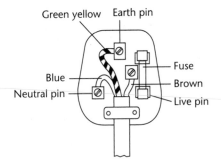

A three-pin plug is now commonly used to connect appliances to the 'mains' by pushing it into a socket connected to the ring main. The wires should be firmly connected to the pins shown in the figure. The yellow/green wire goes to the large earth pin, the b**L**ue wire to the neutral pin on the **L**eft and the b**R**own wire to the live pin on the **R**ight. It is important that *all* of these are correct. The **fuse** should be the smallest possible value for the appliance. The cable grip should hold the insulation on the main part of the cable so that it cannot work loose if it is pulled. The earth pin is longer than the others so that it can open the safety shutters in the socket and so that the safety wire is connected first.

-∔- **Domestic electricity, Earth, Fuse**

POLES

✦ *Magnetic fields, Magnetic poles*

POTENTIAL DIFFERENCE (P.D.)

The potential difference between two points is the energy produced by each unit of electric charge that passes between the two points.

1 *volt* is the potential difference between two points when 1 joule of **energy** is produced by each coulomb of **charge** flowing between them.

We can therefore find the energy that changes state when charge flows around any part of the circuit from

$$\text{energy changed} = \text{p.d.} \times \text{charge moved}$$
$$= V \times Q$$

and the energy will be in joules if the p.d. is in volts and the charge is in coulombs.

This is the same as:

$$\text{p.d.} = \frac{\text{work done}}{\text{charge moved}}$$

Also since

$$\text{charge} = \text{current} \times \text{time}$$
$$\text{energy changed} = V \times I \times t$$

(see also **Heating effect of electric current**).

The total of all the potential differences in a **series** circuit will be the same as the terminal voltage of the supply. The e.m.f. of the supply will be the terminal voltage plus the internal voltage that is needed to drive the current through the supply and complete the circuit. (See also **Electromotive force**.)

In order to measure a p.d. you need to connect a *voltmeter* across the two points, i.e. in parallel with the circuit. This means that the voltmeter must have a very high resistance, or a current will use it to by-pass part of the circuit and may even change what is happening in the circuit.

The idea of potential always causes problems for students when they first meet it. If you are still not happy about it try thinking of the potential as equivalent to pressure in a piped water system. Just as the water flows from high pressure to low pressure, the current flows from high potential to low potential. As the water flows from higher to lower pressure it releases energy, which can drive turbines. As the current flows from higher to lower potential it also releases energy that can be turned into such forms of energy as heat and light. The pump that drives the water is the source of both the pressure and the energy. The battery in the circuit is the source of both the potential and the energy.

> Remember: The earth is at zero potential (0 V). Higher potentials will try to drive current to this zero potential, which is why there is danger of electric shock as 'mains' voltages try to drive current through you to earth.

> Remember: It is the CURRENT that moves and the p.d. that makes it go. Do not make nonsense statements in exams about voltages flowing around circuits.

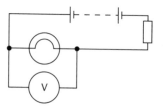

(a) Using a voltmeter to measure the p.d. across a bulb

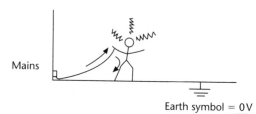

Mains

Earth symbol = 0 V

(b) Electric shock as current is driven from high potential to low potential

✦ *Charge, Current, Electromotive force, Heating effect of electric current, Potential divider, Power*

POTENTIAL DIVIDER

In a circuit each component has a **potential difference** across it that depends on its **resistance** and the **current** flowing through it. If you have two resistors that are in **series**, the current is the same in both and the p.d. depends only on their resistance. We can use this idea to divide up voltages to get the ones that we need.

An example calculation may help.

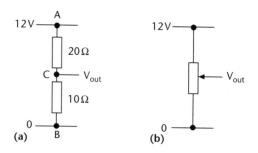

In figure (a):

$$\text{total resistance} = R_1 + R_2$$
$$= 20 + 10 = 30\,\Omega$$

$$\text{current in resistor} = \frac{V}{R} \text{ (Ohm's law)}$$

$$= \frac{12}{30} = \frac{2}{5} \text{ A}$$

$$\text{p.d. across BC} = IR = \frac{2 \times 10}{5} = 4\,\text{V}$$

So the voltage at C is 4 V. Notice that the output voltage is a third of the supply voltage and that the resistance of BC is also a third of the total. In figure (b) you will get 4 V only if the wiper of the **potentiometer** is a third of the way along. You would get 0 V if it was at the bottom and 12 V if it was at the top. As the wiper is moved along you can get any voltage output that you like between 0 and 12 V.

Notice that the values work out in this way only if no current is taken at V_{out}, because the calculation assumed that the same current went through both resistors. In a real circuit this is not much of a problem as long as you make sure that the current that leaves at V_{out} is small compared with the current through ACB.

In the 'perfect' case, when no current leaves along V_{out}, the output voltage can be found from

$$V_{out} = V_{in} \times \frac{R_2}{(R_1 + R_2)}$$

There are several sensors that change their resistance under different conditions (e.g. **thermistor, LDR**) and these can be put into a potential divider so that they produce an output voltage that depends on their resistance.

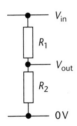

CHECKPOINT

An LDR of resistance 5 kΩ is connected in series with a 25 kΩ resistor across a 9 V supply. What is the p.d. across the LDR?

-|- **LDR, Ohm's law, Resistor, Thermistor**

POTENTIAL ENERGY

Potential energy is the **energy** that an object has because of its position or its condition.

If you lift an object upwards, you do **work** against gravity and that same quantity of energy can be recovered by allowing the object to return to its original level. The additional energy that the object contained at the higher position is called **gravitational potential energy.**

If you stretch a spring, you do work against the elastic forces of the spring. It too will contain some additional energy, which is released when the spring is allowed back to its original length. In a similar way, stretched elastic, a bent beam, a wound-up spring and a taut bow all contain some potential energy, which can quickly be transformed into other forms of energy.

Two magnets that are close to each other have the ability to do work as they attract or repel and have magnetic potential energy. Two electric charges will also have electrical potential energy when they are close together, and the forces between them can be used to do work.

-|- **Energy, Gravitational potential energy**

POTENTIOMETER

This is a **resistor** with a terminal at each end and a sliding contact that can be moved along it, called a wiper. The type used in electronics has a carbon track and is often called a 'pot'. The resistor can be made from wire so that it can be used with higher currents and is then called a *rheostat*.

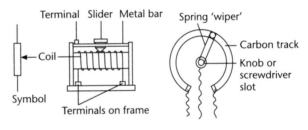

(a) Rheostat   **(b) Potentiometer**

If only one end and the wiper are connected to a circuit, the potentiometer can be used as a variable resistor. Its other main use is as a **potential divider**.

-|- **Potential divider, Resistor**

POWER

Power is the rate at which **work** is done, *or* is the rate of change of **energy** from one form to another.

Work done depends only on the force and the distance moved and the time taken to do it is not taken into account. We are often concerned about how long it takes to do the work and this makes the measurement of power important.

1 watt (W) is the **power** when 1 joule of work is done each second.

As long as we measure the work in joules and the time in seconds, we can find the power in watts from:

$$\text{power} = \frac{\text{work done}}{\text{time taken}} \qquad P = \frac{W}{t}$$

or

$$\text{power} = \frac{\text{energy changed}}{\text{time taken}}$$

Remember: The amount of work done is the same as the amount of energy that changes state, so energy is also measured in joules.

Electric power uses the same unit, the watt, and

power = p.d. in volts × current in amps
$$P = VI$$

Worked example
A kettle is marked 2,400 W. How much heat energy will it supply in 3 minutes?

$$\text{power} = \frac{\text{energy changed}}{\text{time taken}}$$

$$2,400 = \frac{\text{energy changed}}{3 \times 60}$$

energy changed = 2,400 × 3 × 60 = 432,000 J
heat produced = 432 kJ

CHECKPOINT

A lift carries a 2,000 N load through a vertical height of 20 m in 1 minute. What is the power of the lift?

✛ **Energy, Heating effect of electric current, work**

PRESSURE

Pressure is the normal force on a unit area of surface.
A normal **force** is one acting along a normal, i.e. at 90° to the surface. This force is quite often the weight of a body that is resting on the surface. If we know the total force and the total area it acts on, then

$$\text{pressure} = \frac{\text{normal force}}{\text{area}} \qquad P = \frac{E}{A}$$

This gives the pressure in N/m², and we use this so often that it has its own name: 1 N/m² is called 1 **pascal** (Pa).

Worked example
A machine weights 500 N and rests on a base 0.5 m square. What pressure does it produce on the floor?

area on which force acts = 0.5 × 0.5 = 0.25 m²

$$\text{pressure} = \frac{\text{normal force}}{\text{area}}$$

$$= \frac{500}{0.25} = 2,000 \text{ Pa}$$

CHECKPOINT

Why does your thumb have so little effect if you press on a notice board when a drawing pin goes in easily?

Pressure caused by a liquid

The pressure caused by a **liquid** increases uniformly with depth. This means that if you double the distance below the surface of the liquid you will find that the liquid has doubled its pressure. The wall of a dam is thicker at the bottom so that it can withstand a greater pressure. The pressure also depends on the **density** of the liquid. Mercury is 13.6 times as dense as water and will have 13.6 times the pressure of water at the same depth. In many practical problems you will have to remember that the air will exert an additional pressure on the surface of the liquid.

Thicker base to withstand greater pressure

pressure caused by a liquid = depth × density × g
$$P = hDg$$

(g = gravitational field strength = acceleration due to gravity).
Atmospheric pressure is often stated as 760 mm Hg, which means that it is the same pressure as is caused by a 760 mm depth of mercury. You can use the equation above to show that this is the same as 101,000 N/m².

Worked example
What is the pressure caused by a 10 m depth of water? (Density of water = 1,000 kg/m³, g = 10 N/kg.)

$$\text{pressure} = \text{depth} \times \text{density} \times g$$
$$= 10 \times 1,000 \times 10$$
$$= 100,000 \text{ Pa}$$

It is interesting that this is close to the value for atmospheric pressure and that a diver has therefore doubled the pressure on his body at this depth, since the air pressure is also pushing down on the surface!

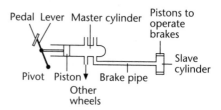

The **molecules** of a liquid are very close together, so a liquid cannot be squashed into a smaller space, but the molecules are still free to move about. This will mean that if you apply a pressure to one part of a liquid, the same pressure will be felt in all parts of it. This is used in car brakes to transmit the pressure from the foot pedal to liquid in a 'master cylinder' and from there to a 'slave cylinder' at each wheel (see figure). The pistons in the slave cylinder are then pushed outwards to operate the brake.

Pressure of a gas

The pressure of a **gas** is not caused by its weight but

by its molecules bumping into the surface (see **Kinetic theory**). A lot of molecules hitting the surface quickly will, however, cause a large pressure. This is true of atmospheric pressure, which is the same as the pressure caused by a 10 m depth of water or 760 mm depth of mercury. Pressure of gases is often quoted as an equivalent depth of liquid in this way. Since mercury has a large density it is used for large pressures, as it gives a reasonable depth that we can imagine easily. The pressure of the atmosphere is measured using a **barometer**. The pressure of many other gases is measured using a **manometer**.

Several experiments show that the atmospheric pressure is large at the Earth's surface.

1. A metal can fitted with a bung and a thick-walled rubber tube can have its air pumped out by a vacuum pump. The pressure outside is then much bigger than the pressure inside and the can is squashed. A simple hand pump is good enough to do this – you do not need to pump out all the air with an expensive electric pump.

2. A more traditional experiment was first done by the mayor of Magdeburg in 1654. A smaller version is often used in school demonstrations. The air is pumped out from the inside of a sphere made from two closely fitting halves. It is then impossible to pull them apart because of the air pressure pushing them back together. After air is allowed back in through the valve they separate easily because the pressure is the same inside as out.

-♦- **Barometer, Force, Manometer**

PRESSURE LAW

If a fixed **mass** of **gas** is kept at a constant volume, its pressure is directly proportional to its absolute temperature.

If the mass of gas is fixed, there will always be the same number of molecules in the container. If the temperature rises, the molecules move more rapidly and hit the walls harder (see **Kinetic theory**). This will raise the pressure. If the gas is cooled until its molecules stop (we cannot really do this, because the gas turns to liquid and then solid – but it is a useful idea!), the pressure would be zero as there would be no collisions with the walls. This temperature is 'absolute zero' and you cannot make a material any colder. The increase in pressure is directly proportional to the increase in temperature, provided that we measure the temperature from absolute zero using the kelvin **temperature scale**.

$$\frac{\text{pressure}}{\text{temperature}} = \text{constant}$$

$$= \frac{P_1}{T_1} = \frac{P_2}{T_2}$$

where P_1 = first pressure, T_1 = first temperature, P_2 = second pressure, T_2 = second temperature. Temperature *must* be in kelvin (add 273 to temperature in °C).

Worked example

An 'empty' can contains air at a pressure of 1 atmos. It is thrown on a fire, where its temperature increases from 20 °C to 1,500 °C. What is the pressure of the air inside at this new temperature?

The volume is constant, so we can use the pressure law:

$$\frac{P_1}{T_1} = \frac{P_2}{T_2}$$

$$\frac{1}{293} = \frac{P_2}{1,773}$$

$$P_2 = \frac{1,773 \times 1}{293}$$

$$= 6 \text{ atmos}$$

Check your own syllabus carefully to find out if you need to know all this work for exams. Some syllabuses only require you to be able to *describe* how temperature and pressure affect each other and leave the problem solving to the **general gas equation**.

-♦- **Boyle's law, Charles' law, General gas equation, Kinetic theory**

PRIMARY COLOURS

Addition

When two or more beams of coloured light are reflected from the same part of a white screen we see their combined effects as a new colour. Three colours cannot be made in this way, and these are the *primary* colours: red, blue and green.

Adding together the primary colours gives secondary colours.

Primary colours		Secondary colour
red + blue	→	magenta
red + green	→	yellow
blue + green	→	cyan
red + blue + green	→	white

A colour TV works in the same way, illuminating tiny spots of red, blue and green to make all the colours that we see.

Subtraction

The pigments in paints and dyes work by absorbing most of the colours that fall on them and reflecting the colours that we see. Mixing two of these will subtract more colours, so that the result is different from colour addition. The primaries in this case are red, blue and yellow. For example:

$$\text{blue} + \text{yellow} \rightarrow \text{green}$$

-♦- **Electromagnetic spectrum, Spectrum**

87

PRINCIPAL AXIS

-ꞏꞏ- *Lens*

PRINCIPAL FOCUS

-ꞏꞏ- *Lens*

PRINCIPLE OF MOMENTS

-ꞏꞏ- *Moment*

PROJECTOR

A projector is used to put an image of a slide or film onto a screen. The slide is first illuminated so that it can become the object for the projection lens. It is placed between F and 2F of the projection lens, so the image produced on the screen is magnified, real and inverted (see **Lens**).

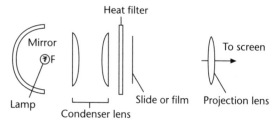

The parts of a projector

The bulb is usually a quartz–iodine bulb, which gives a lot of light but less heat than a filament lamp. If an ordinary filament bulb is used there will probably be a fan to keep the film cool. The concave mirror makes sure that most of the light goes forward, and the condenser lenses, with their special shape, illuminate the slide evenly. The heat filter is there to absorb heat from the lamp, which might damage the slide.

The two condenser lenses make sure that the slide is lit evenly. If these lenses are not there the middle of the picture is brighter than the rest. The lenses are flat on one side and convex on the other.

Moving the projection lens away from the slide will make a picture that is smaller and closer to the projector. The lens is usually on a screw mount so that it can be moved in this way and focus the picture on the screen at the distance that you need.

-ꞏꞏ- *Lens*

PROTON

A proton is one of the three particles from which atoms are made. It has a positive charge and a mass of one unit. It is found in the nucleus of the atom.

-ꞏꞏ- *Atom, Electron, Neutron*

PROTON NUMBER

The proton number is the number of **protons** in the nucleus of an atom. All atoms of the same element have the same proton number.

-ꞏꞏ- *Atom*

PYROMETER

-ꞏꞏ- *Thermometer*

QUALITY OF A MUSICAL NOTE

The pitch of a musical note depends on its *frequency* – the higher the frequency the higher the pitch. This frequency is called the fundamental. Mixed in with this fundamental frequency will be other frequencies, which give the note its quality (timbre) and make the same note sound different on different instruments, even though it has the same fundamental frequency. A multiple of the fundamental frequency is called a harmonic, and the additional frequencies that are mixed with the fundamental frequency will be harmonics. Not all the harmonics will be present and those that are will vary in intensity from one instrument to another. Those harmonics that are present are called the overtones.

Amplitude, Frequency, Sound

RADIATION OF HEAT

✦ **Infrared radiation**

RADIO WAVES

✦ **Electromagnetic spectrum**

RADIOACTIVE DECAY

Sometimes the nucleus of an **atom** is not quite stable, because the balance of its **protons** and **neutrons** is not quite right. The nucleus will then decay by emitting a particle or some electromagnetic radiation. There is no way to predict when this will happen for a particular atom, because the event is entirely random, but we can be quite accurate about what will happen when there are lots of atoms (see **Half life**).

Experiments show that three different types of radiation can be emitted. These are called **alpha**, **beta** and **gamma**. Their symbols are α, β, γ. All three types come from the *nucleus* of the atom and not from the surrounding electrons. They can be shown to be quite different by the effect of a strong magnetic field.

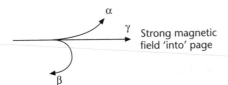

Strong magnetic field 'into' page

The field required to deflect alpha particles is *much* stronger than that required for beta particles. It is not possible to deflect gamma radiation. This suggests that alpha particles have a + charge (see **Fleming's left-hand rule**) and beta have a − charge. Gamma radiation is not deflected at all and has no charge. It is made up of electromagnetic waves of very short wavelength. All three types will create **ions** in the material that they travel through and can be detected by a **Geiger counter**.

There will always be a background of natural radiation that comes from radioactive materials in the Earth's crust or is reaching us from space. This varies considerably from place to place, depending mainly on the type of the local rocks – even the stone or bricks from which buildings are made can emit some radioactivity. If you are doing experiments that involve using a Geiger counter to measure count rates, you will have to measure the *background count* and subtract it from your readings. You can do this by standing the counter on the bench, well away from

radioactive materials, and finding the count rate over several minutes. You can then divide the count by the time to find the counts per second. It is best to do this at the start and end of the experiment and take an average.

All three types can be dangerous and should be treated with great caution. They are not safe just because you cannot see them. The safety precautions for each type depend on their particular properties, which are given in their descriptions, but there are some rules that apply to them all:

1. Count out all sources for an experiment. Count them back in at the end. If the source is a powder or a liquid it should be weighed at the start and the end of the experiment. All radioactive materials must be accounted for.

2. Each source should be kept in its own lead-lined container labelled with its name and always kept in the same locked safe in a room with the door clearly labelled.

Radiation warning sign

3. All sources should be as small as is practical for their use.

4. All sources should only be handled with a pair of handling tongs that keeps them at a safe distance from the body. *Never* try to look into a sealed source by getting your eyes close to the source.

5. All three types of radiation become *much* more dangerous if the source is inside the body; and great care must be taken to ensure that no radioactive dust, powder or liquid is inhaled or eaten – including 'small' quantities from fingers.

Effects on people

These depend on the type of radiation and on the dose received. The ions produced as the radiation passes through tissue can destroy or change cells. A high dose will kill the cells or at least stop them multiplying properly. This may lead to sickness, skin burns, loss of hair or death as the organs of the body stop working properly. A lower dose can change the way that the cells work, resulting in cancer, leukaemia, and hereditary defects that can be passed on to children.

Uses

The uses for radioactive nuclides are mainly of the following types:

Thickness measurement and control

This can be carried out with a beta source on one side of the material and a detector on the other. If the thickness changes, more or fewer beta particles reach the detector, and a computer, noting the change in count, can automatically readjust the thickness. This sort of control is used in the production of paper to keep the thickness constant.

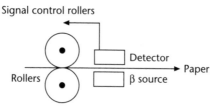

Paper production

Sterilization

The sterilization of hospital instruments and other equipment can be carried out by killing the bacteria with a large dose of gamma radiation. This can be especially useful with materials that would be damaged by heat sterilization. Food can be sterilized before packing in the same way so that it remains 'fresh'.

Medical

Medical uses include directing beams of gamma radiation at cancer cells so that the cancer cells are killed. Patients can be given small samples of radioactive elements or chemicals. These will have a short half life so that they are not dangerous but can be traced by the gamma radiation that they give out to see how the body deals with those particular chemicals.

Gamma rays

Gamma rays can be used in a similar manner to X-rays to detect cracks in large metal castings or welds. They can be used for greater thicknesses of metal than X-rays, and the equipment is more portable.

Tracer experiments

Tracer experiments can be done by adding small quantities of radioactive material to existing chemicals and following where they go with a Geiger counter. This technique can have many different applications. Adding a radioactive chemical to the waste leaving a pipe out to sea will enable engineers to find out where the waste goes. Making plant nutrients slightly radioactive enables biologists to find out where the nutrients go in plants.

Carbon dating

Carbon dating can be done by measuring the quantity of radioactive carbon-14 left in materials that were once alive (wood, wool, etc.). All living materials are based on carbon atoms and contain a certain proportion of carbon-14. After the material dies, the proportion of carbon-14 in it slowly decreases, with a half life of 5,570 years. If you know how much is left you can estimate the time since the material died.

Most of the useful radioactive isotopes are made by exposing a suitable stable nuclide to neutrons in a nuclear reactor. The result will be an unstable nuclide of the same element. Since it is chemically identical to the original nuclide it is ideal for use as a tracer, etc.

✦ *Alpha particles, Beta particles, Gamma radiation, Geiger counter, Half life, Fission, Nuclear energy*

REAL IMAGE

A real image is one that can be projected onto a screen. The rays of light actually pass through the image and do not just appear to come from it.

✦ *Curved mirror, Lens, Virtual image, Reflection*

RECTIFIER

Many applications require **direct current** rather than **alternating current** and a rectifier circuit is used to change the readily available a.c. (from the 'mains' and a **transformer**) into d.c. In its simplest form the rectifier is a diode, so that current can only flow round the circuit one way. This produces a **half-wave rectifier**. This is wasteful, because half the a.c. supplied is not used, and a **full-wave rectifier** can produce a much better result.

✦ *Half-wave rectifier, Full-wave rectifier*

RED SHIFT

When a source of light moves away from you, the wavelength of that light becomes longer. This is called a red shift because red is at the longer-wavelength end of the spectrum.

The **spectrum** from a **star** that is moving away from us will be moved towards the red end of the spectrum. The faster the star moves away the bigger the shift – this allows scientists to work out how fast galaxies and stars are moving.

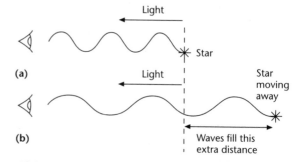

If the star is not moving away from us we see the light as in (a). If the star moves away while it is emitting the light then the waves are spread over a greater distance and have a longer wavelength as in (b).

A **star** emits a continuous spectrum from its hot interior. As the light passes through the outer gases of the star the atoms of the gas absorb the wavelengths of light that are the spectrum for their element. The spectrum that reaches us will therefore have a pattern of dark lines that corresponds to the elements in that particular star. It is called an *absorption spectrum*. We know exactly where these patterns should be for each element and can therefore measure how much the pattern is shifted towards the red. We can then work out the speed of the star or galaxy that emitted the light.

The results show that the stars and galaxies that are farthest from us are also moving away from us fastest.

✦ **Big Bang theory, Galaxy, Star, Universe**

REFLECTION

When **waves** hit a hard surface they will usually be reflected. Each narrow straight beam, called a ray, will be reflected so that:

1. The angle of incidence = the angle of reflection.

2. The angle of incidence and the angle of reflection both lie on the same plane.

Normal

i   r

Remember that both angles are measured from the normal, which is a line at 90° to the surface of the reflector. In the diagram, *i* is the angle of incidence and *r* is the angle of reflection. You can check that all this is true using a strip of plane mirror and a ray box to reproduce the rays in the diagram on a piece of plain paper. To make diagrams simple, mirrors are drawn as a line with lots of small 'hatch' lines on the *back* of the mirror to show the silvering.

(a) Reflection from a polished surface   **(b) Diffuse reflection**

If the reflector is highly polished and flat, it will reflect parallel rays as parallel rays. A mirror is a good example of this. If the reflector is not flat, the incident rays are scattered at different angles, giving *diffuse* reflection like that from a piece of white paper.

When light rays from an object reflect from a plane mirror the reflected rays will produce an image. We

can work out where this must be from the laws of reflection. The eye will see an image of the object that will always obey the following rules:

1. A line from the image to the object will cross the mirror at 90°.

2. The object and the image are both the same distance from the mirror.

3. The image is a **virtual image**.

4. The image is the same size as the object. It sometimes *seems* to be smaller because it is further away!

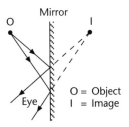

O = Object
I = Image

5. The image is *laterally inverted* (see diagram). This means that it is turned sideways but not upside down.

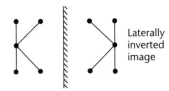

Laterally inverted image

You should check these rules by drawing some images for yourself – perhaps your initials are suitable. Remember to draw the plane mirror with the symbol that has been used in these diagrams.

> *Remember: Applying these rules to a more complicated shape shows that the image will always obey the rules.*

✦ **Curved mirrors, Ripple tank, Virtual image, Wave**

REFRACTION

Refraction is a change of the direction in which a **wave** is moving and is caused by the change in **velocity** as the wave moves from one material into another.

All waves have this property, which can easily be demonstrated with a **ripple tank**. The most common uses are for changing the direction of rays of light as it goes in and out of different materials such as glass or plastic lenses.

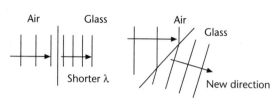

(a) Waves along normal   **(b) Waves refracted at boundary**

The diagrams show light going from air into glass and therefore slowing down. Since the frequency must stay the same (waves are not disappearing!), the wavelength gets shorter. If the waves hit the boundary at 90° (along a normal) their direction does not change. If the waves hit the new material at an angle, one edge of the beam is slowed down before the other edge and the beam is turned as in the diagram above. The beam will, of course, be turned in the opposite direction when it leaves the material. This is seen when light goes through a rectangular glass block and emerges parallel to the direction in which it entered (see diagram).

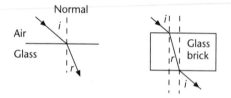

The angle of incidence is the angle between the ray going towards the boundary and the normal. The angle of refraction is the angle between the ray leaving the boundary and the normal.

> *Remember: Measure the angles from the normal and NOT from the boundary.*

Light will be refracted towards the normal when it enters a denser material, so the angle of incidence *i* is bigger than the angle of refraction *r*. It will be refracted away from the normal when it leaves the more dense material.

It is easy to check these diagrams, and measure the angles, using a glass block and the narrow beam of light from a ray box. If you put the glass block on plain paper, draw around it and send in the ray from the ray box, you can mark the position of the rays with pencil dots. Lift the apparatus off the paper and draw in the complete path of the ray with your pencil and ruler. If you are going to measure the angles remember that they are measured *from the normal*, so you will need to draw the normals first!

In addition to use in lenses, etc., refraction also causes the effect known as *apparent depth*.

Sound can also be refracted by sending it through gases of different density so that the speed is changed.

✦ *Apparent depth, Critical angle, Lens, Ripple tank, Wave*

REFRIGERATOR

A refrigerator works by pumping heat from the inside to the outside. The cabinet must be lined with a good insulator (plastic foam) and have a good door seal (airtight magnetic strip all the way round) so that the heat cannot re-enter easily and the machine is reasonably efficient.

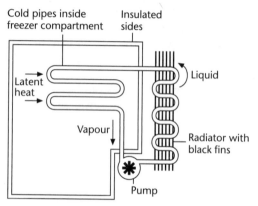

A refrigerator

The liquid is volatile, and the reduced pressure caused by the pump causes it to vaporize in the pipes inside the fridge. This removes *latent heat* from the inside of the cabinet, which is cooled. When the liquid reaches the radiator at the rear of the cabinet it is being compressed by the pump and condenses back into a liquid. This releases the latent heat and the heat is lost into the surrounding air by the radiator. The pump can be turned on and off by a *bimetallic* thermostat inside the cabinet.

✦ *Latent heat*

RELAY

A relay is an electromagnetic switch that enables a large current to be switched by a small current, or a large voltage to be switched by a small voltage.

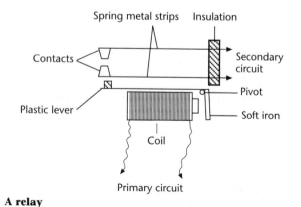

A relay

A small current in the primary circuit flows through the coil and it magnetizes the core. The core attracts the soft iron, pulling the lever so that the contacts are pushed together. The secondary circuit has then been turned on. If the primary circuit is turned off, the core no longer attracts the iron on the lever and the springy contacts separate, turning the secondary circuit off. A relay like this is said to be *n*ormally *o*pen (NO).

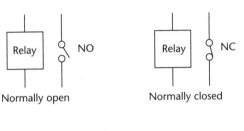

Normally open   Normally closed

Change over

In some relays, the contacts are pushed apart by the plastic lever (instead of together) and the secondary circuit is on until the primary circuit turns it off. This sort of relay is *n*ormally *c*losed (NC).

In other relays, the lever switches the contacts in the secondary circuit from one setting to another. This type is called a *c*hange *o*ver relay (CO).

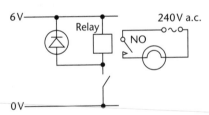

When the current in the coil is switched off, a large voltage can be induced (see **Electromagnetic induction**). This could damage other components in the circuit, especially **transistors**. A diode is placed in parallel with the coil so that the current produced flows through it without causing problems (see diagram).

⊹ **Magnetic poles and magnetic forces, Transistor switch**

RESIDUAL CURRENT CIRCUIT BREAKER

⊹ **Domestic electricity, Earth**

RESISTANCE

Resistance is a measure of how difficult it is for **current** to flow through part of a circuit.

A part of a circuit has a resistance of 1 *ohm* if a p.d. of 1 volt drives a current of 1 amp through it. The symbol for ohms is Ω. To work out resistances you will need **Ohm's law**. You will also find experiments on resistance under that heading.

The resistance of a wire will increase with the length of the wire and decrease with a larger cross-sectional area. It will also depend on the metal that the wire is made from.

If you are investigating this you may find it useful to know that:

$$\text{resistance} = \frac{\text{resistivity} \times \text{length}}{\text{area}}$$

Resistance will be in Ω, length will be in m, and the area, which is the cross-sectional area of the wire, will be in m^2. Resistivity is a constant that depends on the metal that the wire is made from and is measured in Ωm.

A path in the current can flow from one terminal of a supply to the other with almost no resistance is called a *short circuit*. This must be avoided, as the large current produced will damage the supply.

⊹ **Heating effect, Ohm's law, Resistor**

RESISTOR

A resistor is an electrical/electronic component that is designed to have a particular resistance. Its value will change if it is not operated at the correct temperature.

Some resistors are made from wire, but most small fixed resistors for electronic circuits are made from carbon and are colour-coded by a series of 'bands' to tell you the value and tolerance. (The tolerance is the maximum error that is allowed by the manufacturer, and most resistors are much closer to their stated value than this might lead you to expect – a 10 per cent tolerance is quite good enough for most electronic circuits.) You will not need to remember this colour code for examinations, but it is included with some examples because it is so useful in practical work and projects.

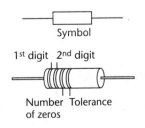

RESISTOR COLOUR CODES

first band = first digit   second band = second digit
third band = number of zeros   fourth band = tolerance

Colour	Meaning	Colour	Meaning
black	0	green	5
brown	1	blue	6
red	2	purple	7
orange	3	grey	8
yellow	4	white	9

fourth band
gold = ± 5%   silver = ± 10%   no band = 20%

Worked example

A resistor with brown-, black-, red- and silver-coloured bands has the value 1,000 Ω and a 10 per cent tolerance, i.e. 1 kΩ ±10%.

CHECKPOINT

What is the value of a resistor that is marked with yellow, purple, orange and gold-coloured bands?

Resistors in series

Since the current will have to flow through all resistors that are placed in **series**, their values are added together.

$$R_{total} = R_1 + R_2 + \ldots$$

R_1 R_2 R_3

$10\,\Omega$ $20\,\Omega$ $30\,\Omega$

The three resistors illustrated will therefore have a total resistance of $10 + 20 + 30 = 60\Omega$.

CHECKPOINT

What is the total resistance of this pair of resistors?

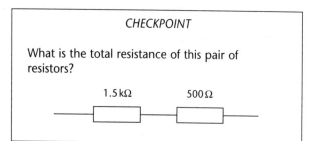

$1.5\,k\Omega$ $500\,\Omega$

Resistors in parallel

If you imagine one of the resistors on its own, its resistance is a measure of how difficult it is for current to pass. Fitting in another resistor in **parallel** provides another route for the current, so the total resistance must be lower. This will mean that the total resistance must be less than the resistance of the smallest resistor. There is a formula to help you work out the value:

$$\frac{1}{R_{total}} = \frac{1}{R_1} + \frac{1}{R_2}$$

R_1

R_2

$12\,\Omega$

$6\,\Omega$

In the example illustrated:

$$\frac{1}{R_{total}} = \frac{1}{R_1} + \frac{1}{R_2} = \frac{1}{12} + \frac{1}{6} = \frac{1}{4}$$

$$R_{total} = 4$$

Remember: The most common error in doing these problems is in forgetting to do the last line, so that you give an answer for $1/R$ instead of R.

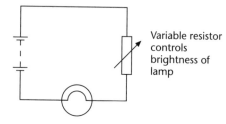

Variable resistor controls brightness of lamp

A variable resistor can be used to change the current in a circuit (see figure). If you make its resistance greater, less current flows around the circuit.

✦ *Current, Parallel, Resistance, Series*

RESONANCE

An object can vibrate with its own natural **frequency**. If it receives vibrations from outside at that frequency then it will begin to vibrate with a large **amplitude**. This is called resonance.

It can be a nuisance when the engine vibration of a lorry makes its driving mirrors vibrate, but it is useful when the tuning circuit of a radio resonates at the frequency of a radio signal picked up by the aerial.

A wind instrument contains a column of air that resonates at a particular frequency. Different lengths of that column produce different notes. In a similar way, stringed instruments have strings that resonate. In this case, the frequency depends on the length, tension and mass of the string.

✦ *Amplitude, Frequency*

RESULTANT FORCES

A resultant force is the single force that will replace several separate forces – a sort of total.

The directions of the forces are important as well as their sizes and can be allowed for (see **Vector**).

If forces are parallel (acting in the same direction or in opposite directions), you can find the resultant by adding those in the same direction and subtracting those in the opposite direction.

CHECKPOINT

Find the resultant force in each of the small diagrams:

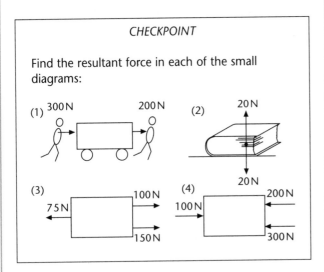

 Vector

RHEOSTAT

 Potentiometer

RIGHT-HAND GRIP RULE

When a **current** flows along a **conductor** it produces a **magnetic field** in the space surrounding it. The field is circular, with the wire at its centre. This rule tells you which way the field goes around the wire.

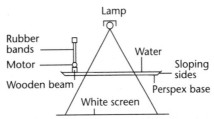

Wire with current 'into page'

Wire with current 'out of page'

Imagine that you grip with your right hand so that the thumb points in the direction of the current. The fingers then go around the wire in the same direction as the field. Note that on this sort of diagram current towards you is shown as a dot in the circle representing the wire. Current away from you is a cross.

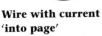 **Magnetic field, Electric motor effect**

RIPPLE TANK

A ripple tank shows what happens to water ripples as they are reflected or refracted at different surfaces. Since these are general properties of waves, the patterns produced also apply to other types of **wave**.

The tank is shallow and has a perspex base with sloping sides.

A ripple tank

As a ripple crosses the surface, the light above the tank shows the ripple as a shadow on the screen below. (You can see this sort of shadow pattern on the bottom of an ordinary sink or bath – if you dip a finger in and out of the water you can see circular shadows of the ripples.) Straight (plane) waves can be produced by dipping a short wooden beam in and out of the water. This is usually done by hanging the beam on two rubber bands and making it vibrate up and down by means of a small electric motor driving an eccentric. If circular ripples are required they are produced by raising the bar and fitting a small plastic ball to it so that the ball just touches the surface. The sloping sides stop the waves being reflected from the sides and complicating the pattern.

The patterns can be made to appear stationary by viewing them through a spinning disc with a slot cut in it. If the disc spins at the correct speed the waves will have moved on by one **wavelength** each time the slit comes around, so the pattern always looks the same. This is the 'stroboscopic effect'. Another method is to replace the lamp with a flashing light (stroboscope) and adjust the flash rate until it is the same as that of the motor on the wooden beam.

Strips of aluminium about 30 mm wide can be bent into the appropriate shape and used as mirrors by standing them in the path of the waves in the tank. You will produce patterns that show that the laws of **refraction** and the images produced by **curved mirrors** are all properties of waves.

Refraction depends on the speed of the waves, and water ripples can be slowed down by making them travel in a smaller depth of water above a piece of perspex. The patterns produced are similar to those shown in the section on refraction.

Diffraction and **interference** patterns can also be seen by allowing water ripples to pass through narrow gaps in barriers in the tank.

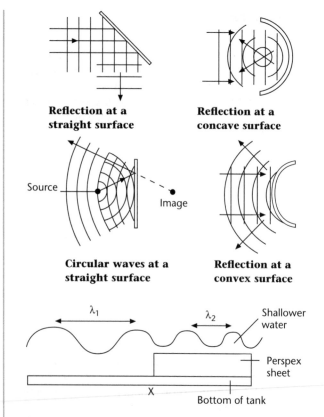

Reflection at a straight surface	**Reflection at a concave surface**
Circular waves at a straight surface	**Reflection at a convex surface**

The change in depth at X causes the wave to slow down so that $\lambda_2 < \lambda_1$ and refraction occurs

-⊹- **Diffraction, Interference, Reflection, Refraction, Wave**

R.M.S. VALUES

-⊹- **Alternating current**

ROCKET

Fuel and oxygen are carried in tanks inside a rocket. They are burned together inside the combustion chamber and produce a large quantity of hot (fast moving) gas molecules. These leave the engine at the rear and the engine is pushed forward by an equal 'reaction' force. Since the engine carries all that it needs it can work anywhere, including outside the atmosphere.

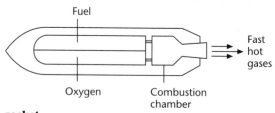

A rocket

A simple illustration of the principle behind the engine can be made by making a small but fairly heavy car and using a balloon of air to make it move.

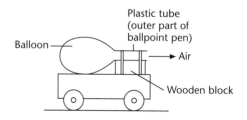

-⊹- **Jet engine, Newton's third law**

ROGUE RESULTS

Very few people can do an experiment without one or more results that do not fit the pattern of the other results. These are the rogue results and should be checked by repeating that part of the experiment. They often prove to be the result of a mistake by the experimenter and can then be ignored. They are much easier to spot on a graph than in a table of results. Do remember, however, that many important discoveries would have been missed if the results had not been checked rather than just ignored.

> *Remember: If you find that you have a rogue result do not include it in your averages or plot it on your graph. Do include it in your table and discuss why it happened in the evaluation part of your experiment and say why you left it out. Do try to repeat that result and get a better value.*

-⊹- **Graph**

SATELLITE

A satellite is a body that rotates around a planet in an orbit. It may be man-made like a television satellite, or natural like a moon. The satellite and the planet attract each other (the satellite is in the **gravitational field** of the planet, and this is the **centripetal force** needed to keep it in orbit).

There are two types of man-made satellite.

● *Polar orbital satellite:* This orbits around the Earth passing over both poles. As the Earth is also rotating on its axis the satellite will be over a different part of the Earth on each orbit. After a few orbits, it will have passed above all the Earth's surface. It can be used to watch clouds and weather and track hurricanes. Other uses include watching military movements and checking on crops and climate.

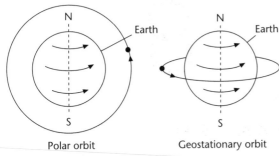

Polar orbit   Geostationary orbit

Orbits for satellites (not to scale)

● *Geostationary satellite:* This orbits directly above the Equator (but very high), completing exactly one orbit each day. The Earth below is also rotating once each day, so the satellite appears to stay in the same position above the surface. This makes it very useful for relaying telephone calls and beaming down TV programmes.

Satellites are also used to explore the solar system and look at the stars. Pictures from Earth are always made poorer by the atmosphere and the dust in it, which scatter the light. Pictures from space do not have this problem and are much sharper and clearer. Some other radiation from space is absorbed by the atmosphere and can only be observed properly from satellites. This includes research about the age of the universe.

-**⚹**- *Solar system*

SCALAR

A scalar quantity is one that has a size but no direction. Scalar quantities of the same type may simply be added to find a total.

   Mass is a scalar and so is ***speed***.

-**⚹**- *Vector*

SECOND

-**⚹**- *Time*

SECONDARY COLOUR

A secondary colour is produced by the mixing of two **primary colours**.

-**⚹**- *Primary colours*

SEMICONDUCTOR

This is a type of material that is often used to make electronic devices such as ***transistors, diodes, integrated circuits***, etc. The two that are used are silicon and germanium, which are insulators in their pure state. They are made into better conductors by 'doping' them by adding tiny quantities of special impurities. One type of impurity increases the number of electrons in the crystal and makes an n-type semiconductor. A different impurity leaves some gaps in the electron structure, which behave like positive holes. This is a p-type material.

Devices made from semiconductors will be small and reliable but cannot take large voltages such as our 'mains' voltage, a maximum being about 35 to 40 V. If too much current is driven through them they become hot, which makes them have a smaller resistance and take more current. This process is called thermal runaway and rapidly damages the device. You must connect any semiconductor device to a heat sink if it carries much current.

-**⚹**- ***Diode, Integrated circuit, LED, Transistor***

SERIES

Components are in series in a circuit when the **current** must flow through each of them in turn. This means that the current does not divide along different routes in that part of the circuit.

The first diagram shows several bulbs, a battery and a switch connected in series. The same current will flow through all the bulbs, and the supply voltage will need to be the sum of all the voltages needed for the individual bulbs. If one of the bulbs fails, all the bulbs go out, because the circuit is not complete and no current can flow round it. An

Planet	Mass (Earth = 1)	Diameter (Earth = 1)	Surface Gravity (Earth = 1)	Density (Water = 1)	Distance from Sun A.U.	Composition	Time for one orbit (years)	Time for one rotation on axis	Moons
Mercury	0.06	0.38	0.38	5.4	0.39	rocky	0.24	58.7 d	0
Venus	0.82	0.95	0.9	5.3	0.72	rocky	0.62	24.3 d	0
Earth	1	1	1	5.5	1	rocky	1	24 h	1
Mars	0.11	0.53	0.38	3.9	1.52	rocky	1.88	24.6 h	2
Jupiter	320	11	2.7	1.3	5.2	mainly icy with icy/ rocky core	11.9	9.8 h	16
Saturn	95	9	12	0.7	9.5	mainly icy with icy/ rocky core	29.5	10.2 h	20+
Uranus	14.5	4	1.2	1.2	19.2	icy/rocky	84	17.9 h	15
Neptune	17	4	1.2	1.7	30	icy/rocky	164.8	19.1 h	2
Pluto	0.003	0.18	0.2?	2	29	icy/rocky	248	6.4 d	1

1 A.U. = 1 astronomical unit = average distance from Sun to Earth.
The table has been made so that it is easy to compare the planets with Earth. If you need 'real' figures, the following facts about the Earth will let you work them out:
Mass = 6×10^{24} kg; Diameter = 12,800 km; Surface gravity = 9.8 N/kg.

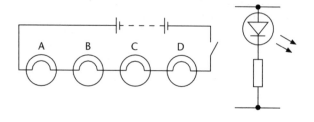

example of this sort of circuit is in Christmas tree lights, where twenty 12 V bulbs are connected in series across the 240 V 'mains' supply.

The second diagram shows a **resistor** and an **LED** in series.

> *Remember: Cells can be connected in series to form a battery, and their voltages are added. A 9 V battery for your transistor radio will contain six 1.5 V cells connected in series.*

 **Current, Parallel, Resistors**

SEVEN-SEGMENT DISPLAY

This is an output device for electronic circuits that can display numbers and letters by turning on some or all of the segments. The individual segments will be either **LEDs** or liquid crystal, so they can be switched by a small voltage. They are used in watches, calculators, alarm clocks, radios, etc.

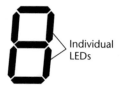

Individual LEDs

✛ **LED**

SOLAR ENERGY

Solar energy can be used to heat water in solar panels. In some 'hot' countries mirrors have been used to focus enough energy in one place to create a solar furnace. The energy can also be used to produce a small electric current from a solar cell. The energy per m² in this country is not sufficient to make large-scale use practical, and our weather makes it unreliable, but it does have some small-scale applications in electronics, where the energy required is small.

✛ **Sources of energy**

SOLAR SYSTEM

This consists of nine planets, which revolve around a central star, the Sun, in orbits. There is also a band of smaller objects, called asteroids in an orbit between Mars and Jupiter. Some of the planets have satellites (moons).

The table above shows some of the properties of these planets, in order of their distance from the Sun.

✛ **Satellite, Sun**

SOLENOID

-+- *Magnetic poles and magnetic forces*

SOLIDS

Solids have all their particles arranged in a regular repeating pattern called a 'crystal lattice'. Each particle has its own place within this structure and is not allowed to leave it. The only movement that is possible for the particle is a random vibration about its position in the lattice. The shape of the solid therefore fixed and it is not possible to compress it into a much smaller space, because the particles are already very close together. The particles are held together by attracting forces, and if you try to stretch a solid you can feel the attracting forces trying to pull it back into shape (see *Hooke's law*).

Putting heat energy into a solid causes the particles to move faster and to have a larger vibration. The particles are using the *specific heat* to increase their kinetic energy and the 'internal energy' of the solid. Eventually, the vibration breaks down the crystal structure and the solid begins to melt. Energy is needed to break down the structure during heating, and we notice this as *latent heat* of fusion. At this point, the temperature remains constant because the energy is being used to pull apart the crystal structure against the attracting forces of the particles instead of making the particles move even faster. The total internal energy of the *liquid* will be much greater than that of the solid because of the latent heat that has been put in. If the liquid is cooled and freezes it will release this same quantity of latent heat.

-+- *Gases, Kinetic theory, Liquids*

SOUND

Sound is a *longitudinal wave* and therefore travels through a material as a series of compressions. It cannot travel through a vacuum, because there are no particles in a vacuum that could vibrate and carry the energy. (See the section on longitudinal waves under *wave*.)

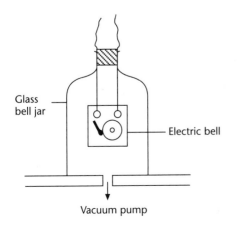

Glass bell jar

Electric bell

Vacuum pump

When the air is pumped out of the jar in the figure, you can see the bell working but you cannot hear it. Men on the Moon needed radios unless their helmets touched!

Since sound is a vibration of the layers of air, its source is always a vibration at the same *frequency*. Sometimes the vibrations are too fast to see easily, especially if the *amplitude* is small (and therefore produces a quiet sound). In other cases, you may be able to see the vibrations if you look closely – e.g. a loudspeaker making a loud low-frequency sound, or the string of a guitar. Your vocal chords in your throat vibrate to produce the sounds for speech. We can hear sounds in an approximate range of 20 Hz to 18 kHz, but this varies considerably from one person to another.

The speed of the sound wave will depend on the material that it travels through and the temperature of that material. The table gives an approximate idea of the speed in some different materials.

Material	Speed in m/s	Material	Speed in m/s
air	330	water	1,500
glass	5,000	brass	3,500
aluminium	5,100	wood	3,850

To measure the speed of sound in air we use the fact that we can see things happen almost instantly, because the speed of light is so fast that the time taken for it to reach you is *very* short. If a pupil stands at one end of a sports pitch and bangs two pieces of wood together, you can measure the time between seeing it happen and hearing it at the other end of the pitch. Measure the length of the pitch and you can find the speed of sound. This is not a very accurate experiment, because the times are so short and difficult to measure. It can be improved by doing it in both directions to allow for any wind. If you have a tall building or a cliff and a lot of space in front of it, you can obtain a much better answer by making a sound and finding the time taken for the echo to return,

To determine the speed of sound accurately you can use a short sound from a small loudspeaker. The loudspeaker is in one side of a container of the material that you are testing and the sound reflects back from the other side. If you pick up the original sound and its echo using a microphone and put the signal into an oscilloscope you can produce a trace with a small peak for each signal. Measuring the distance between them and knowing the time for the spot to cross the screen enables you to work out the time taken for the echo to return, Then

$$\text{speed of sound} = \frac{\text{length of container}}{\text{time taken}} \times 2$$

Sound echoes are often used to determine the depth of the sea or to find shoals of fish. These often use *ultrasonic* sound (above the range of frequencies that we can hear), because they are easier to keep in narrow beams and not confused with engine noises, etc. Ultrasound is often used in medical 'scans'. It reflects from bone and tissue without causing the damage that can be possible with X-rays.

Worked example

A foghorn on a ship sends out a sound and the echo from a cliff is heard 4 s later. How far away is the cliff? (Speed of sound in air = 330 m/s)

$$\text{distance} = \text{speed} \times \text{time}$$
$$= 330 \times 4 = 1{,}320 \text{ m}$$

This is the distance to the cliff and back, so the distance to the cliff = 660 m.

CHECKPOINT

Why do we hear thunder later than we see the lightning flash?

⊹ Amplitude, Frequency, Longitudinal, Wave

SOURCES OF ENERGY

There are a number of major sources of energy. Some, mainly those that are rich in chemical energy, have been used as fuels for a long time. Others are now being investigated as the original fuels run out and become more expensive. Some may have the advantage of being cleaner to produce or of creating less pollution when they are used.

Coal, oil, gas and nuclear

These fuels are all *non-renewable* resources. All are, or will become, more difficult to obtain as we use up existing stocks and have to find new supplies. Coal, gas and oil are known as 'fossil fuels' because of the way in which they were originally made and are burned to release some of their **chemical energy** as heat. Their main disadvantage is the production of air pollution, including carbon dioxide, which increases the **greenhouse effect**. Nuclear power is much cleaner to use and causes fewer problems with air pollution. Some of the waste products, however, will be radioactive for thousands of years (see **Nuclear energy, Nuclear reactor**).

Renewable resources

These can be put into two main groups, **solar energy, wind power, wave power, hydroelectric power** and **biomass**, which all derive their energy from the Sun; and **tidal energy** and **geothermal energy**, which do not.

The following summary table may help you to learn which is which:

Energy source	Renewable	Non-renewable	Energy from Sun originally	Fossil fuel
Gas	×	✓	✓	✓
Coal	×	✓	✓	✓
Oil	×	✓	✓	✓
Electricity	✓	×	(see note)	×
Wind	✓	×	✓	×
Wave	✓	×	✓	×
Tide	✓	×	×	×
Geothermal	✓	×	×	×
Biomass	✓	×	✓	×
Hydroelectric	✓	×	✓	×
Solar	✓	×	✓	×
Nuclear	×	✓	×	×

Note: Electricity *may* have energy that originally came from the Sun, but it depends on which fuel was used to produce it. Electricity is called a secondary fuel because it is produced from another, primary source.

CHECKPOINT

(1) Name the fossil fuels.
(2) Name two fuels that do not derive their energy from the Sun.

⊹ Biomass, Energy, Geothermal energy, Hydroelectric power, Solar energy, Tidal energy, Wave power, Wind power

SPECIFIC HEAT CAPACITY

The specific heat capacity of a material is the quantity of heat energy that will raise the temperature of 1 kg of the material by 1 K without changing its state.

Since the energy goes 'inside' the body, we say that the heat energy has become *internal energy*. The specific heat capacity will be measured in J/kg K.

The definition means that you can calculate the energy needed to change the temperature of a material, provided that it stays solid, liquid or gas. If it changes state it will need extra energy to do so, which is called **latent heat**.

When heat is put into a body its particles gain kinetic energy and we notice the temperature rise as they move more rapidly (see **Kinetic theory**). The heat needed to change the speed of the molecules is different for each material, so each material has a different specific heat capacity. If a liquid is being used to cool something or to store heat, it should have a large specific heat capacity.

$$\text{energy (heat)} = \text{mass} \times \text{specific heat capacity}$$
$$\times \text{change in temperature}$$
$$E = ms\Delta\theta$$

Worked example

1. How much energy is needed to raise the temperature of 2 kg of water from 20°C to 100°C? Specific heat capacity of water = 4,200 J/kg K.

energy = mass × specific heat capacity × change in temperature

$$= 2 \times 4,200 \times 80$$
$$= 672,000 \text{ J}$$
$$= 672 \text{ kJ}$$

2. How long would this take in a 2.4 kW electric kettle?

$$\text{power} = \frac{\text{energy changed}}{\text{time taken}}$$

$$2,400 = \frac{672,000}{\text{time}}$$

$$\text{time taken} = \frac{672,000}{2,400}$$

$$= 280\text{s} = 4 \text{ min } 40 \text{ s}$$

CHECKPOINT

(1) A block of copper absorbs 37,800 J of heat from an electric heater. If its mass is 0.5 kg and its temperature rises from 20 °C to 200 °C, what is the specific heat capacity of copper?

(2) How much energy is released by a 30 kg block of steel that cools from 500 °C to 20 °C? (Specific heat capacity of steel = 500 J/kg K.)

The first checkpoint calculation gives you some idea of how to measure a specific heat capacity. You measure the mass of the material and its temperature and then heat it, using an electric heater. Finally measure its new temperature. The energy put in can be found from the *VIt* equation (see **Heating effect of electric current**). You would need to avoid losing heat to the surroundings as far as is possible. Solids can be wrapped in layers of insulating felt or expanded polystyrene, and liquids can be heated in a vacuum flask.

-⊹- *Kinetic theory, Latent heat, Temperature*

SPECIFIC LATENT HEAT OF FUSION

-⊹- *Latent heat*

SPECIFIC LATENT HEAT OF VAPORIZATION

-⊹- *Latent heat*

SPECTRUM

White light is actually a mixture of waves of many different frequencies, each **frequency** producing a slightly different colour at the retina of the eye. We see this mixture as white. If we split a light source into its separate colours, the result is a spectrum. In a

rainbow the sunlight is split into a spectrum as it passes through the droplets of water.

The light that our eye can detect is only a small part of the **electromagnetic spectrum**. The lower frequencies (longer wavelengths) that we can see appear as red and the higher frequencies (short wavelengths) are blue/violet. You should be able to put the colours of the visible spectrum into their correct order:

red, orange, yellow, green, blue, indigo, violet

increasing frequency
shorter wavelength

The names describe only approximate areas of the spectrum and there are no sudden changes, the colours changing gradually from one to another.

The different wavelengths of light are refracted by different amounts as they pass in and out of materials such as glass, water, and plastics. We can use this to split the light into its spectrum in a process called *dispersion*.

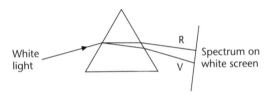

The diagram shows how this can be done to produce a simple spectrum of the light from an ordinary bulb. You could do a similar experiment with a narrow beam of sunlight to find out if both are the same.

Remember: The light is refracted twice and is dispersed a little more each time.

The red, longer wavelength end of the spectrum is always deviated least. With a little patience a better spectrum can be produced from a wider beam of light by putting a cylindrical lens before the prism to focus the colours into a series of lines on the screen. The result is then brighter and clearer.

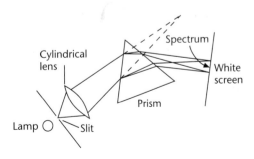

The spectrums emitted by different elements, especially in a gas, are different. Looking at the spectrum from a small sample of vapour can help you to analyse the source material – a larger proportion of an element will make its particular spectrum brighter. This can be done automatically from small samples in the steel industry. The pure spectrum needed is usually produced from a *diffraction grating* rather than a prism.

The visible light is only a part of the spectrum produced, and you can get an indication of this from your simple spectrum. Putting an infrared detector or a phototransistor in the space before the red light should produce a reading on the scale connected to it, showing that the infrared has also been dispersed into its appropriate place. Putting a small quantity of washing powder in the space past the violet end of the spectrum may cause the powder to emit some blue light as it is hit by some ultraviolet. (This may need a darkened room, as the bulb does not emit much ultraviolet, but a beam of sunlight gives better results!)

-+- **Electromagnetic spectrum, Frequency, Wave**

SPEED

Speed is the rate of change of distance with time.

Since no direction is included in the definition, speed is a **scalar**, unlike **velocity**, which is a **vector**. If the speed is uniform (constant) you can find it from

$$\text{speed} = \frac{\text{distance moved}}{\text{time taken}} \qquad v = \frac{s}{t}$$

If the distance is in m and time is in s, the units for speed will be m/s.

If the speed is not uniform and the object **accelerates** or decelerates the equation will give only an average speed. In this case you would get a better answer for the speed at a particular time by drawing a **distance/time graph**.

CHECKPOINT

A cyclist travels 225 m in 15 seconds. What is the average speed in this time?

-+- **Distance/time graph, Velocity**

SPLIT-RING COMMUTATOR

-+- **Electric motor effect**

STABILITY

-+- **Equilibrium**

STAR

A star is a bright, hot object out in space that is sending out the radiation that we can see (other

objects such as planets only reflect the light to us). There are huge numbers of stars, and they are collected together into galaxies.

The **Sun** is our nearest star. The next nearest, *Proxima Centauri*, is about 270,000 times farther away. The distances between stars are very large. We usually measure them in light years, where one light year is the distance travelled by light in one year at a speed of 300,000,000 m/s.

Stars are mostly made from hydrogen and helium and they use these as a fuel in a **fusion reaction** to release a lot of energy.

Some of the stars in the sky seem to make patterns, even though the stars are often enormous distances apart. Many of these patterns have names and are called **constellations**. Finding these patterns can help to identify individual stars. If you look towards the north you might be able to see the constellation called the Plough (Americans call it the Big Dipper, and its correct name is *Ursa Major*). You can find the pole star by looking along a line through the last two stars of the blade of the plough.

Stars seem to move rather than being fixed in the sky. There are two reasons for this:

- *The Earth's rotation*. As the Earth spins on its axis the stars appear to revolve around the axis once every 24 hours. The pole star, *Polaris*, appears to be stationary because it is along the line of the axis. If you look towards the north in the night sky and then do it again a few hours later you will be able to see that the stars have moved around quite a lot (the full rotation would take 24 hours, so 3 hours will see the star pattern rotate by 45° around the pole star).

- *The Earth's orbit*. The stars also change position during the year because we are looking at them from different places on the Earth's orbit around the Sun.

-+- **Constellation, Galaxy, Life cycle of a star, Milky Way, Sun**

STATE

Materials will be in one of three states depending on temperature. These are solid, liquid and gas.

-+- **Change of state, Kinetic theory**

STATIC ELECTRICITY

Static electric **charge** is a fundamental property of **protons** and **electrons**. Equal numbers of each particle seem to have charges that 'cancel' and become neutral with no overall charge. We show this best by letting the charge on a proton be +1 unit and the charge on an electron be −1 unit, so that equal numbers of each have a total charge of 0. The charge on these particles is *very* small (1.6×10^{-19} **coulombs**).

Sometimes a neutral object can lose or gain some electrons so that it has an overall electrostatic charge. If this happens on a conductor, the charge will leak away rapidly, but on an insulator it remains trapped. Take a plastic ballpoint pen and rub it on your pullover sleeve or something woollen. You should find that it can now attract and pick up small pieces of paper. The pen has gained some electrons and a negative charge. The wool will have lost the electrons and be left with an overall positive charge. Rubbing polythene with fur or wool moves electrons from the fur to the polythene; the polythene becomes negative and the fur becomes positive. Rub acetate sheet with a cloth and the acetate becomes positive because it loses electrons to the cloth.

> *Remember: The negative electrons from the outside of atoms move, but the positive protons always stay where they are.*

- negative = gained some electrons
- positive = lost some electrons
- neutral = has equal numbers of protons and electrons

Suspending two pieces of charged material next to each other also shows that:

- like charges repel
- unlike charges attract

Small pieces of dust will be attracted to a charged object such as a record by the same forces that attracted the small piece of paper to your pen. The electrons are repelled to the far side of the dust particle, which is left with a positive charge closer to the negative record. The opposite (unlike) charges attract and the dust 'sticks' to the record. This effect has uses in

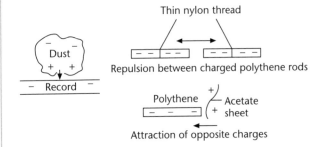

Repulsion between charged polythene rods

Attraction of opposite charges

electrostatic precipitation, which removes the dust from power station chimneys. It has a less useful effect in collecting dust whenever you clean plastic materials by rubbing them!

Charges exert forces on other charges in the space around them. This is an example of a *field*. The strength of this field decreases rapidly with distance as do **magnetic fields** and **gravitational fields**.

CHECKPOINT

A spray that is used for putting insecticide on plants produces a fine spray of charged drops. Explain how this improves the spraying.

✦ *Charge, Current*

SUN

The Sun is made of hydrogen and helium gas; the temperature is about 6,000 °C at its surface. At the centre the temperature is about 13 million °C, and this is where *nuclear fusion* is taking place as the hydrogen is converted to helium. The Sun radiates energy at a rate of 4×10^{26} W. The Sun is about half-way through its life cycle of 9,600 million years and is about a million times larger than the Earth.

The Sun is one of billions of stars that make up a galaxy known as the Milky Way, and there are billions of galaxies like the Milky Way in the universe.

✦ *Solar system*

TEMPERATURE

Temperature is usually thought of as a measure of how hot a body is. **Kinetic theory** tells you that it is really a measure of the speed of the molecules in the body, a higher temperature meaning faster molecules.

It is important to realize that heat and temperature are connected, but that temperature is *not* a measure of heat. Imagine using an electric heater to put heat energy into 200 ml of water in a beaker and measuring the rise in temperature. You could then repeat the experiment with 1,000 ml of water in a larger beaker, switching the same heater on for the same length of time. The water has gained the same amount of heat energy but will not have the same rise in temperature. The same sort of thing will happen if you use the same mass of a different liquid (see **Specific heat capacity**).

⋆ **Kinetic theory, Specific heat capacity, Temperature scale**

TEMPERATURE SCALES

To fix a temperature scale you need two fixed points. The *lower fixed point* is the temperature at which pure ice melts (impurities would lower this temperature). This is called 0 °C on the Celsius scale.

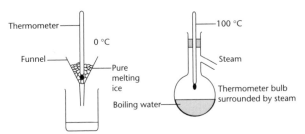

Checking the lower fixed point

Checking the upper fixed point

The *upper fixed point* is the temperature at which pure water boils under a pressure of 760 mmHg (1 atmosphere). This is called 100 °C on the Celsius scale. The range between the two fixed points is divided into 100 equal degrees. A thermometer can be marked with a scale (calibrated) or checked at these points using the apparatus shown in the figures.

A **thermometer** does *not* have to be a mercury-in-glass type that works by expansion of the liquid. Any property that changes reasonably uniformly with temperature will do, and the thermometer is calibrated in the same sort of way.

The real zero of temperature must be the point at which the molecules stop. If they cannot move any more slowly, the temperature cannot go any lower (see **Kinetic theory**). This temperature is called *absolute zero*. On the **kelvin scale** this temperature is zero K and on the Celsius scale it is −273 °C. A change in temperature of 1 K is the same as a change in temperature of 1 °C – they are the same size. To convert from °C to K add 273.

CHECKPOINT

Convert the following temperatures from celsius to kelvin:

0 °C,   100 °C,   27 °C,   20 °C,   −10 °C,   273 °C

⋆ *Temperature, Thermometer*

TENSION

This is a stretching or pulling force, e.g. pulling on a rope or spring.

⋆ *Compression, Hooke's law*

TERMINAL VELOCITY

Terminal velocity is the speed that is reached when the driving force pushing an object in one direction is exactly balanced by the counter forces that push against it. Balanced forces always make an object stay at rest or move at constant velocity; e.g. a cyclist will reach terminal velocity when the forward force from pedalling is balanced by the friction and air resistance forces.

When an object falls in a gas or liquid (such as a sky diver in air or a marble in a jar of oil) it has two forces acting on it. It is accelerated downwards by gravity and slowed by the drag of the fluid through which it is passing. As it falls faster the drag force increases until it is equal and opposite to the weight. The **acceleration** then stops and the object falls at a uniform **velocity**, called terminal velocity.

A sky diver can change the terminal velocity by altering his/her shape and therefore the air resistance but may still have a terminal velocity of 55 m/s! A parachute will produce a greater air resistance and therefore the equilibrium between the forces is achieved at a lower terminal velocity.

The apparatus illustrated may suggest a suitable investigation for coursework. You can use a stopwatch to find the time taken for the marble to fall distance *d*. You should then be able to work out its velocity. Do this several times as a check. Is it really the terminal velocity or do you need a longer tube? Is the answer the same for marbles of different size, mass or density?

LONGMAN **HOMEWORK HANDBOOKS** – **PHYSICS**

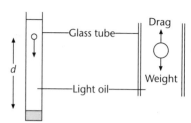

Investigation suggestion

Would it be the same in a different oil? Could you use water? Remember:

● Only investigate one change at a time.

● Put your results into neat, clear, tables with units.

● Draw sensible conclusions from your results and not what you think should happen!

⟊ *Velocity, Weight*

THERMISTOR

The thermistor is a **semiconductor** device that changes its **resistance** as the temperature changes. The resistance gets smaller as it gets warmer. This change can be quite large, but the actual values depend on the type chosen.

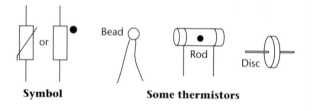

The shape and size can vary from a small bead to a rod 10 mm or so long. A larger thermistor will obviously take more heat to change its temperature and might be unsuitable as the sensor for a thermometer, but would be stronger and cope with bigger currents.

There are many uses for these devices, including thermometers and sensors to switch heaters or cooling fans. Most of these circuits react to a voltage that changes with temperature, and we can obtain this if we put the thermistor in a **potential divider** circuit.

The arrangement shown in the diagram will have a voltage output that goes up as the thermistor gets colder. (It then has a larger resistance and gets a larger share of the voltage applied.)

If you want a circuit that produces an output voltage that increases as the temperature rises, you can swap the two components so that the thermistor is in the upper half of the divider. Whether you do this or not, the circuit will have a fixed output voltage at a particular temperature, and this may not be suitable in your design. The 'cure' is to use a variable resistor instead of the fixed one. You will

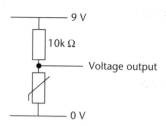

then be able to adjust the output so that it triggers the next stage properly.

This final version is often used as a *temperature sensor* for triggering **logic gates** and **transistor** switches. A thermistor is not suitable for use at high voltages and would need to control 'mains' equipment through a relay.

⟊ *Logic gates, Potential divider, LDR, Thermometer, Transistor switch, Resistance*

THERMOMETER

A thermometer is a device that is used to measure **temperature**.

There are a number of different types of thermometer, which work in different ways and are suitable for different purposes. All depend on some property that changes as the temperature changes.

The liquid-in-glass thermometer

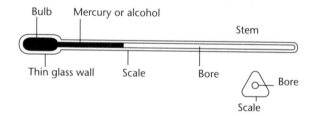

As the liquid gets warmer it expands along the bore of the stem, and the scale is read at the end of the 'thread' of liquid. The bulb of the thermometer contains most of the liquid and has thin walls so that the heat is conducted to the liquid quickly. The bore must be the same width all the way along the stem so that the liquid expands the same distance along the bore for each degree of temperature rise. Making a thermometer with a narrower bore makes the liquid move more for each degree and produces a more sensitive thermometer that may be marked in fractions of a degree. The glass of the stem is often thicker on one side, and this acts as a magnifying glass to enable the scale and liquid thread to be seen clearly. Two different liquids are commonly used depending on the intended use: mercury and alcohol.

Alcohol has a lower freezing point (78 °C) and can be used at lower temperatures. Mercury has a higher boiling point (357 °C) and can be used at higher temperatures. An alcohol thermometer would be broken as the liquid boiled if you tried to measure the boiling point of water, so mercury thermometers are generally used in school/college laboratories. Alcohol thermometers are cheaper to make and are used in greenhouses and tropical fish tanks, where the temperature range is more suitable. Both liquids are dangerous, but mercury is more poisonous, especially the vapour.

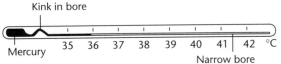

A clinical thermometer

A clinical thermometer is a mercury-in-glass thermometer with a kink in the bore that prevents the mercury returning to the bulb as it cools. After reading the thermometer is shaken to return the mercury past the kink. The bore will be narrow enough to allow temperatures to be read to 0.2 °C but can be short because the temperature scale need only read between about 35 °C and 42 °C (body temperature is 37 °C).

The thermocouple

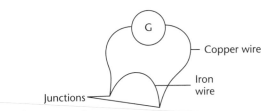

A thermocouple

The thermocouple has two wires of different metals joined as shown in the diagram. If the two junctions are at different temperatures a voltage is produced and a reading is obtained on the galvanometer. A larger temperature difference produces a larger reading. The scale of the meter is usually marked in degrees so that the temperature can be read directly. The junctions can then be made very small and strong and take in very little heat, which is often useful. The thermocouple can be used at much higher temperatures than a liquid-in-glass thermometer.

The thermistor

A *thermistor* has an electrical resistance that decreases as the temperature increases. This can be used as a temperature sensor if it is fitted into a potential divider, as shown in the figure. As the temperature increases so does the reading on the

meter. The device can be made very small and sensitive but will be damaged by temperatures above about 50 °C.

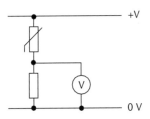

The pyrometer

In the case of very hot materials (furnaces, etc.) an ordinary thermometer would melt. A pyrometer works by looking at the colour of the light that is given out by the hot object.

-╬- *Temperature scale*

TICKER TIMER

A ticker timer is a small device that makes fifty dots each second on a tape that is pulled through it. This leaves a record on the tape of the motion of the body pulling the tape. A uniform **velocity** will produce a pattern of equally spaced dots. **Acceleration** is shown by a steadily increasing distance between dots. Since the time between dots is known and the distance moved can be measured on the tape with a ruler, you can work out the velocity or acceleration.

-╬- *Newton's second law, Weight*

TIDAL ENERGY

Tides are caused by the gravitational attraction of the Moon and the Sun, which attracts the water in the oceans into two slight bulges on the surface of the Earth.

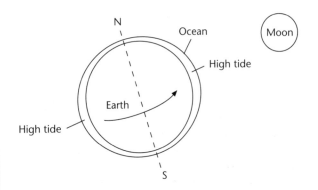

As the Earth spins once each day, places on the surface pass through these two higher water levels, so we have two tides each day. At each tide, the water level rises and then falls again. We can use this by making the tide flow through barriers, driving turbines as it does so. The main problems are in providing suitable sites and the large initial building costs,. There are also worries about the ecological effects of the reduced tide after the barrier and the visual impact of the barrier on coastal sites.

-◆- *Sources of energy*

TIME

The unit for time is a fundamental unit called the second.

The second has had several definitions (one based on a fraction of the year 1900), but it is now defined by an 'atomic' clock as the time taken for 9,192,631,770 vibrations of caesium-133 atoms. The definition is an important one, but you will not be required to remember exact numbers like this!

-◆- *Appendix One*

TOTAL INTERNAL REFLECTION

-◆- *Critical angle*

TRANSFORMER

A transformer is a device that is used to change electric voltages. A step-up transformer will increase the voltage and a step-down transformer will make the voltage smaller.

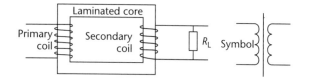

The primary coil carries an **alternating current**, so that the soft iron core is magnetized by a continually changing field (see **Magnetic poles and magnetic forces**). This changing field passes through the secondary coil and an e.m.f. is induced in the secondary coil which obeys **Faraday's law**. If the number of turns on each coil is the same, the voltage on the secondary (output) coil is the same as that on the primary (input) coil. If the number of turns on the secondary coil is larger than that on the primary, the voltage is stepped up. If the number of turns on the secondary coil is less than on the primary, the voltage is stepped down.

> *Remember: A transformer must have an alternating current. Direct current would not produce a CHANGING magnetic field in the core.*

$$\frac{\text{secondary voltage}}{\text{primary voltage}} = \frac{\text{turns on secondary coil}}{\text{turns on primary coil}}$$

or

$$\frac{V_{\text{out}}}{V_{\text{in}}} = \frac{N_{\text{secondary}}}{N_{\text{primary}}}$$

> *Remember: Some people prefer to work out the right-hand side of this equation and call it the 'turns ratio'. If the primary voltage is multiplied by this you have found the secondary voltage. Maths students will see that the two methods are really the same!*

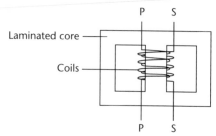

In order to make sure that all of the changing magnetic field passes through the secondary coil, the coils are often wound one on top of the other, as in the diagram. They are, of course, well-insulated from each other. The core will be built up of lots of layers of iron with thin layers of oxide between them as an insulator. This stops currents being induced in the core instead of the coils. Such 'eddy currents' would heat up the core and waste energy. The **efficiency** of a well-designed transformer can be almost 100 per cent, in which case:

power input = power output

$$\frac{\text{primary}}{\text{volts}} \times \frac{\text{primary}}{\text{amps}} = \frac{\text{secondary}}{\text{volts}} \times \frac{\text{secondary}}{\text{amps}}$$

You should remember that transformers will work only on a.c. (d.c. would not produce the changing magnetic field that the device depends on) and therefore your 'mains' electricity must be a.c. (see **National Grid System**). If d.c. is needed, produce the required voltage using a transformer and then use a **rectifier**.

Worked example

A transformer is to produce a 12 V output from a 240 V supply.

1. If the primary winding has 600 turns, how many are required for the secondary coil?

$$\frac{V_{out}}{V_{in}} = \frac{N_{secondary}}{N_{primary}}$$

$$\frac{12}{240} = \frac{N_{secondary}}{600}$$

$$N_{secondary} = \frac{600 \times 12}{240}$$

$$= 30 \text{ turns}$$

2. If the transformer supplies 2 A when it is correctly wired up, what is the input current?

$$\frac{primary}{volts} \times \frac{primary}{amps} = \frac{secondary}{volts} \times \frac{secondary}{amps}$$

$$240 \times primary\ amps = 12 \times 2$$

$$primary\ amps = \frac{12 \times 2}{240}$$

$$= 0.1 \text{ A}$$

(Although the voltage has been stepped down the current has been stepped up by the same factor.)

CHECKPOINT

A small electric welder uses a transformer to obtain a 2 V supply from the mains. Its primary coil has 460 turns. The mains is 230 V a.c. How many turns will there be on the secondary coil?

There are *many* applications for transformers: transistorized equipment such as radios, amplifiers, etc. need low voltage, but our 'mains' supply is high; welding equipment needs high current at low voltage, so a step-down transformer is used; toys such as train sets and Scalectrix need 12 V d.c. from a mains supply, so a step-down transformer is used followed by a **rectifier**.

-**⁌**- *Electromagnetic induction, Faraday's law, National Grid System*

TRANSISTOR

A transistor is a **semiconductor** device with three terminals. The three terminals are called the *base, collector* and *emitter* (see figure).

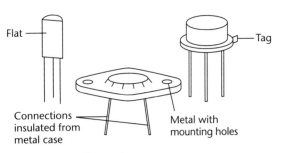

A npn transistor (silicon)

A pnp transistor (germanium)

b = Base
c = Collector
e = Emitter

Transistors come in many shapes and sizes. They will have only three connections and one of those might be the metal case. The only sure identification is from the number on the case.

Flat — Tag

Connections insulated from metal case

Metal with mounting holes

Different types of transistor

Most types are now npn and are based on silicon. A larger transistor will be able to conduct more current without overheating, but a transistor that gets hot in use will probably have to be bolted to a heat sink. (A heat sink is a piece of metal, usually aluminium, that conducts away the heat.) Most transistors will not survive being connected to more than 30 or 40 volts, and very small ones will be able to conduct only small currents. The transistor can be used as a very fast switch or as a current amplifier. To do this it is now often built into an **integrated circuit**, and such a device can contain very large numbers of transistors and other components in a very small space.

-**⁌**- *Integrated circuit, Semiconductor, Transistor switch*

TRANSISTOR SWITCH

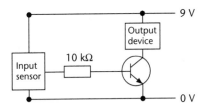

9 V

Output device

10 kΩ

Input sensor

0 V

The input sensor produces a voltage that is connected to the base of the **transistor** through a **resistor** (see diagram). This voltage will switch the transistor on or off and this in turn switches the output device. There are a variety of input and output devices, discussed below.

If the input voltage is low (less than 0.6 V), there will be no 'base current' through the base and emitter to the 0 V line. The resistance through the collector and emitter will be very high, and very little current flows through the transistor from the collector to 0 V. The transistor is *off*. Since there is no current through the output device the voltage across it will be zero and the collector will be at about supply voltage (9 V in the diagram).

If the input voltage is high (more than 0.6 V) a small current flows through the resistor and then the base and emitter to 0 V. This makes the resistance through the collector and emitter low and the current flows through the transistor from 9 V to 0 V. The transistor is now *on*. The collector current flows through the output device and switches it on. Because there is now a voltage across the output device the voltage at the collector falls to almost 0 V.

> Remember: It takes at least 0.6 V to drive a small current through the base–emitter, but that this turns the transistor on and allows a big current to flow through the collector–emitter.

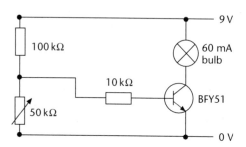

A good way to test this circuit is to build the circuit in the above diagram. There are many ways to do this using special circuit boards that enable you to put the components together quickly without soldering. In this case, the output device is a bulb, which shows clearly when the transistor is on and off. A bicycle dynamo bulb is best, because it will work on a fairly small current (about 60 mA). The input device is a **potential divider** and you can turn the base voltage up and down by adjusting the variable resistor so that you can see the result of the transistor switching.

You are still doing the switching yourself by adjusting the resistor, but it is easy to turn the circuit into one that switches automatically. If you use a

sensor consisting of an **LDR** or a **thermistor** in a potential divider you will get a light-controlled or temperature-controlled switch. See the entries for LDR and thermistor for more details.

A bulb may be unsuitable as the output device and you can use a suitable **relay** or an LED as an alternative. There are many possible combinations of sensor and output devices. Try drawing some circuit diagrams and carefully go through what will happen. Two examples are illustrated.

In (a), the resistance of the LDR increases as it gets darker, and the voltage output from the potential divider therefore increases. When it reaches 0.6 V, it switches the transistor on and current flows through the relay. This turns the 'mains'-operated lamp on. Similar devices are sold as security lights.

In (b), the LED will come on when the thermistor reaches a pre-set temperature. A relay in place of the LED could operate a cooling fan.

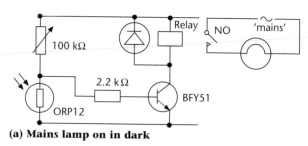

(a) Mains lamp on in dark

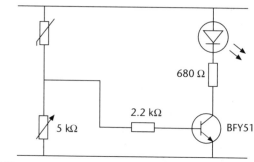

(b) Temperature-controlled switch

⊶ *Potential divider, Thermistor, Transistor*

TRANSVERSE WAVE

This is a type of wave in which the vibration is at right angles to the direction of travel. Water ripples and all electromagnetic waves, such as light, are transverse.

⊶ *Wave*

TRUTH TABLE

⊶ *Logic gates*

ULTRAVIOLET

-+- *Electromagnetic spectrum*

UNIVERSE

This is the collection of all the galaxies in space and time. We believe that the universe is expanding and that other galaxies except those in the 'local group' are moving away from us. The evidence for this is in the **red shift** of the light from other galaxies.

The galaxies that are farthest away from us are also the ones that are moving away from us fastest. This does not mean that we are at the centre of the universe, only that we are part of a whole universe that is expanding outwards as though it had come from a single starting place. This is evidence for the so-called **Big Bang theory**.

-+- **Big Bang theory, Galaxy, Red shift**

U-TUBE MANOMETER

-+- *Manometer*

U-VALUES

A U-value is a quantity used by heating engineers and architects to work out the rate at which heat will be conducted through walls, etc. of buildings.

The U-value of a material is the heat energy lost through each square metre of the material for each 1 °C of temperature difference across the material.

$$\frac{\text{energy}}{\text{loss}} = \text{U-value} \times \frac{\text{surface}}{\text{area}} \times \frac{\text{temperature}}{\text{difference}}$$

Some example U-values are given in the table. They are all in $W/m^2 K$.

Material	U-value
Double-brick cavity wall	1.7
Double-brick wall with cavity insulation	0.6
Single-glazed window	5.6
Double-glazed window 20 mm air gap	3.0
Tiled roof	2.2

CHECKPOINT

A double-brick cavity wall measures 2.5 m high and 4 m long. It has a double-glazed window in it that is 1 m high and 2 m long. The temperature inside is 20 °C and it is 5 °C outside. Use the table of U-values to work out the energy loss through the wall.

-+- **Conduction of heat, Conductor**

VACUUM FLASK

The vacuum flask was originally designed to keep **liquids** cold but is good at stopping heat transfer in or out and can be used equally well to keep things hot.

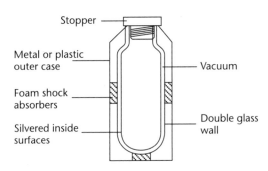

Stopper

Metal or plastic outer case

Foam shock absorbers

Silvered inside surfaces

Vacuum

Double glass wall

A vacuum flask

- **Conduction** is reduced by the vacuum between the glass walls and by the stopper, which is usually plastic with a foam interior.

- **Convection** is also reduced by the surrounding vacuum and the stopper, which prevents convection currents between the contents and the air above the flask.

- **Infrared radiation** is prevented by the silvering of both interior walls of the glass container.

There will be slow heat transfer, caused mainly by conduction through the walls of the glass container at the neck and conduction through the stopper.

-⁙- *Conduction, Convection, Infrared radiation*

VAPOUR

A vapour is formed by the faster-moving molecules of a liquid leaving its surface. These form a vapour in the space above and can exert a pressure by collision with the walls of the container in the same way as a gas. The pressure of the vapour will depend on the temperature of the liquid. A vapour is not the same as a gas, because its pressure does not depend on its volume. If it is compressed into a smaller space the pressure remains constant and some of the vapour condenses back into a liquid.

-⁙- *Kinetic theory*

VECTOR

A vector quantity is one that has a direction *and* size.

- **Velocity** is a vector because it is the speed in a certain direction.

- **Force** is a vector because it will act in a particular direction.

In order to add two vector quantities you will also need to add their directions. It is easier to this if you think of the + as meaning 'followed by'. You can then find the total by either a scale diagram or a calculation. The total is called the resultant.

Worked example

A sack of grain hanging on a rope has a mass of 50 kg. It is being pulled forward by a horizontal force of 300 N. What is the resultant of these two forces? First remember that the problem is about **forces** and *not* **mass**, so you need the **weight** of the sack.

weight of sack = mass × gravitational field strength
$$= 50 \times 10$$
$$= 500 \text{ N}$$

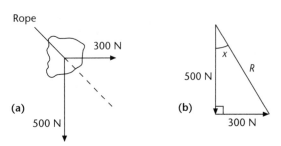

Sketch the resulting forces as in the diagram. Draw one of the forces and then the second force, starting from the end of the first. The resultant is then represented by the line from the start of the first to the end of the second. You can either calculate this or draw all the forces to scale and find the resultant from the length of the line and the angle *x*.

By Pythagoras:
$$R^2 = 500^2 + 300^2$$
$$R^2 = 250{,}000 + 90{,}000$$
$$R^2 = 340{,}000$$
$$R = 583 \text{ N}$$

Also
$$\tan x = \frac{300}{500}$$
$$= 0.600$$
$$x = 31°$$

The resultant force is 583 N at 31° to the vertical.

Do not forget to work out the direction. If you find it easier to do the final parts by scale drawing you will not lose marks in an examination. You should also note that, if the sack is stationary, the forces are in equilibrium, so the tension in the rope must be exactly equal and opposite to the resultant. This method is often used by engineers and is called the 'triangle of forces'.

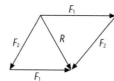

R is the resultant of *F₁* and *F₂*

When the vectors are not at 90° it is sometimes easier to find the resultant by a scale drawing that completes the parallelogram (see diagram) by drawing the two vectors in the opposite order (second followed by first). This will give exactly the same answer but can be drawn more accurately and is easier than calculations in a triangle that does not have a right angle.

⊶ *Force, Resultant forces, Scalar, Velocity, Weight*

VELOCITY

Velocity is the rate of change of **displacement** with time, *or* the rate of change of distance with time in a stated direction, i.e. **speed** in a particular direction. The units will therefore be m/s in a stated direction.
Sometimes the direction is very obvious, on a straight track for example, and we do not write it down, but it should normally be included because velocity is a **vector**, unlike speed.
We can find velocity from

$$\text{velocity} = \frac{\text{change of displacement}}{\text{time taken}}$$

$$\text{Or velocity} = \frac{\text{distance moved in stated direction}}{\text{time taken}}$$

$$v = \frac{s}{t}$$

You can also find velocity from a **distance/time graph** and from a **velocity/time graph**.

⊶ *Displacement, Distance/time graphs, Speed, Vector, Velocity/time graphs*

VELOCITY RATIO

⊶ *Machine*

VELOCITY/TIME GRAPHS

These graphs show the motion of an object more clearly than a table of data and can also be used to find the acceleration or distance moved by the object. At this examination level the acceleration will be constant, so the graphs are straight lines.

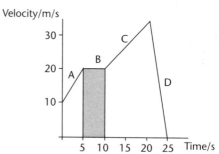

Velocity/time graph for an object moving along a straight track

The graph shows an object that accelerates from 10 m/s to 20 m/s in the first 5 s. It then moves with uniform (constant) velocity for 5 s before accelerating to 35 m/s in the next 10 s. In the final 10 s it decelerates uniformly to a stop. (It does *not* move backwards!) All of this can be read from the scales of the graph. You can now find the **acceleration** in each stage. For example:

$$\text{acceleration in A} = \frac{\text{change of velocity in A}}{\text{time taken}}$$

$$= \frac{20 - 10}{5} = \frac{10}{5} = 2 \text{ m/s}^2$$

If you are good at maths you will have seen that this is finding the *gradient* of the graph.

CHECKPOINT 1

Find the acceleration in each section.

You can also find the **displacement** in each case. Look at part B of the graph. The object moves at 20 m/s for 5 s, so that the displacement in $20 \times 5 = 100$ m along the track. This is also the same as the area of the shaded rectangle under part B. This will always happen – the distance moved is the same as the area under the graph in the time required.

CHECKPOINT 2

Now find the area under the graph for each section so that you know the distances moved, and find the total distance moved.

Remember: You may find it easier to split the areas into rectangles and triangles, find their areas and then add them together.

⊶ *Acceleration, Displacement, Distance/time graphs, Vector, Velocity*

VIRTUAL IMAGE

A virtual image is an image that cannot be found on a screen. Usually this means that the light does not actually pass through it but appears to come from it.

> Remember: This does not mean that the image does NOT really exist – it must exist, because you can see it! This type of image is often produced by both mirrors and lenses.

✛ **Curved mirrors, Lens, Real image, Reflection**

VISIBLE LIGHT WAVES

✛ **Electromagnetic spectrum**

VOLT

The volt is a unit of **potential difference**. If 1 **coulomb** of electric charge flows between two points with a p.d. of 1 volt between them, 1 **joule** of energy is transferred.

✛ **Potential difference**

VOLTAGE

✛ **Volt**

WATT

One watt (W) is the **power** when 1 **joule** of work is done each second.

WAVE

A wave is a regular disturbance, usually a vibration, that transfers **energy** from one place to another. Usually this results in particles vibrating, but it may be a magnetic and/or electric field that vibrates, as in **electromagnetic** waves such as light or X-rays.

There are two general types of wave, depending on the direction in which the particles or field vibrates. Note that in both cases the particles vibrate only about their original rest position and do not travel along with the wave. To describe a wave exactly you would need to note its **wavelength, frequency** and **amplitude**.

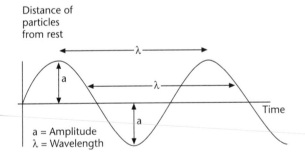

a = Amplitude
λ = Wavelength

Transverse wave

In this type of wave the vibration is at right angles to the direction of travel. Ripples on water are this type of wave, and a **ripple tank** is often used as a demonstration.

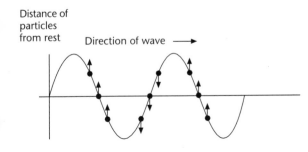

The small arrows in the diagram show the direction of the vibration of the particles at that time. As the wave moves on, the position of each particle and its velocity will be continually changing in a repeating pattern. You can make **transverse** waves by vibrating one end of a piece of rope that is fixed at the other end. Electromagnetic waves are also transverse waves.

Longitudinal waves

In this type of wave the vibration is backwards and forwards parallel to the direction of travel. This causes a series of compressions to travel along.

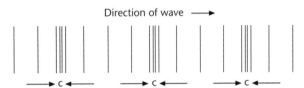

Direction of wave ⟶

The small arrows in the diagram show the direction in which the particles have moved at that time. Some of the particles have been moved together, causing the compressions lettered C. Sound is a **longitudinal** wave. You can also produce longitudinal waves by extending a spring and then pulling together and releasing some of its coils. A 'Slinky' does this very well!

All waves will show a similar set of properties, which include **reflection, refraction, interference** and **diffraction**.

CHECKPOINT

1. Name the two main types of wave. Give an example of each.
2. How do the particles vibrate in each of these two types?
3. What is carried from one place to another by all waves?

⬦ **Amplitude, Diffraction, Electromagnetic spectrum, Frequency, Interference, Reflection, Refraction, Ripple tank, Wave equation, Wavelength**

WAVE EQUATION

The velocity of a **wave** can be found from its **frequency** and its **wavelength** by using the wave equation:

velocity = frequency × wavelength
$$v = f \lambda$$

The **velocity** will be in m/s provided that the frequency is in Hz and the wavelength is in m.

> *Remember: 1,000,000 Hz = 1 MHz and 1,000 Hz = 1 kHz*

If you are comparing two waves of the same type it is sometimes useful to remember that they will travel with the same velocity in the same material, so the wave with a higher frequency must have a shorter wavelength (to keep frequency × wavelength the same).

Worked example

A radio station transmits a signal with a frequency of 100 MHz. If the velocity of the radio waves through the air is 3×10^8 m/s, what is the wavelength of the signal?

$$\text{velocity} = \text{frequency} \times \text{wavelength}$$
$$300{,}000{,}000 = 100{,}000{,}000 \times \text{wavelength}$$
$$\text{wavelength} = \frac{300{,}000{,}000}{100{,}000{,}000} = 3\,\text{m}$$

CHECKPOINT

Jamie measures the frequency and wavelength of a sound wave. She finds that they are 250 MHz and 1.36 m. What is the speed of the wave? She changes the frequency to 200 Hz. What do you think the wavelength will become?

✦ *Frequency, Wave, Wavelength*

WAVELENGTH

Wavelength is the length of one complete *wave*. It is measured from any point on the wave to the next place where the particles are moving in the same direction and are at the same point in their vibration. Particles that are behaving like this are said to be 'in phase'. Wavelength is usually given the symbol λ and is measured in length units (m or mm).

> *Remember: It is common in examinations to be given a diagram of a wave and be asked to measure the wavelength. Take care not to measure just to the next place on the diagram that is the same height above the rest position, because the particle may be going in the opposite direction and you will have measured part of a wavelength.*

✦ *Frequency, Wave*

WAVE POWER

Sea waves often carry large quantities of energy, because a large mass of water is being moved up and down. Exploiting wave power involves strings of machines on the surface of the sea. As the waves pass by they drive part of the machine up and down and this drives a hydraulic pump. The fluid driven by the pump in turn drives a generator. The waves emerge smaller than before encountering the machine. The main problem has been the large

variation in size of the waves and the damage done by bad weather. A lot of machines would be needed and suitable sites could be difficult to find. The original source of this energy is the Sun – not the Moon as in **tidal energy**.

WEIGHT

Weight is a **force** caused by gravity acting on a mass. Any two objects that have mass will be attracted towards each other (even people!), but at least one of the objects must be very massive for the force to be large enough to matter. Since weight is a force it *must* be measured in **newtons**. Each 1 kg of mass at the surface of the Earth will be attracted to the Earth with a force of approximately 10 N, that is, the **gravitational field** strength of the Earth at its surface is approximately 10 N/kg (actually 9.81 N/kg).

$$\text{weight} = \text{mass} \times \text{gravitation field strength}$$
$$W = mg$$

This means that, on Earth, the weight in N of a mass in kg is given by:

$$\text{weight} = \text{mass} \times 10$$

For instance, a person of mass 60 kg will have a weight of 600 N. This is the force of attraction towards the Earth. The Earth is attracted towards the person with the same force, but it is so massive that it is not going to move very far!

Weight is a force and is therefore a **vector**. It is important that you do not confuse weight with **mass** – a weightless object moving in space will still have mass and **momentum**. An astronaut will still feel pain if he hits his thumb with a weightless hammer!

If you find it difficult to imagine the size of 1 N, remember that a bag of sugar from the supermarket will weigh about 10 N. Sir Isaac Newton would also have been pleased to know that an average eating apple weighs about 1 N!

Weight will cause objects to accelerate 'downwards'. This **acceleration** should be the same for *all* objects because a larger force (weight) will have to accelerate a correspondingly larger mass. If you drop a ball and a piece of paper together this seems not to be true, as the paper drifts gently downwards. If you screw the paper up into a ball to reduce its air resistance and try again you can see that they do fall together and that the problem was caused by the extra force of the air resistance. A better version of this experiment was done by an American astronaut on the Moon. He dropped an eagle feather and a hammer at the same time. They could be seen to fall together because there was no air resistance. They also accelerated more slowly because the Moon's smaller mass produces a smaller gravitational force, about one-sixth of that on Earth.

The acceleration (g) caused by the Earth's gravity at its surface varies slightly, depending on where you are on the surface, but it is about 9.81 m/s². For most simple problems we use $g = 10$ m/s². This 'acceleration due to gravity' can be measured by using a falling mass to pull a paper tape through a *ticker timer*.

-+- *Acceleration, Force, Mass, Vector*

WIND POWER

Wind power can be obtained from modern windmills with rotors that are rather like the propellers of an aeroplane in shape. Large rotors can produce enough electricity by driving a generator to be economical, but there are problems with lack of wind – or too much some of the time – and objections about the large unsightly structures in country areas. If these are to provide an alternative energy source on a large scale there would have to be many of them on westward-facing hills, and many people may find the effect on the landscape unacceptable. The original source of energy is the Sun.

-+- *Sources of energy*

WORK

When a *force* moves an object some work must be done. The amount of work will depend on the size of the force and on how far it moves (but *not* on how long it takes).

If a force of 1 *newton* moves a distance of 1 m in the direction of the force, 1 *joule* of work is done.

The joule (J) is the unit for work and for *energy*. The definition means that we can find the work done from the equation:

work done = force × distance moved in the
direction of the force

$$W = Fs$$

Remember to put the force in N and the distance in m to get the work done in J. If the time taken to do the work is important you will need to look up *power*.

> Remember: When work is done, energy is changed from one form to another. We use the same unit, the joule, to measure energy and often think of energy in terms of how much work it can do.

Worked example

A piece of stone weighing 100 N is lifted up 4 m to the top of a building. How much work is done?

In order to lift a weight of 100 N there must be an upward force that is at least 100 N.

work done = force × distance moved
= 100 × 4
= 400 J

CHECKPOINT

A boy of mass 60 kg ran up some stairs until he was 10 m higher than before. How much work was done?

-+- *Energy, Force, Machine, Power*

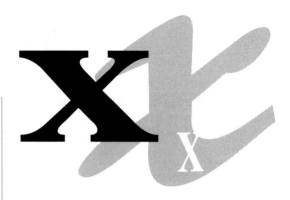

X-RAYS

X-rays are very short-**wavelength** (high-**frequency**) electromagnetic **waves**. They are usually produced by firing a beam of high-speed electrons at a metal such as tungsten. The energy is re-emitted from the electrons of the metal as electromagnetic waves. The waves have sufficient energy to cause cell damage and cancers if received in large doses, but they are quite safe in the small quantities used for medical purposes. Their uses depend on their penetrating power and their effect on a photographic plate.

✦ *Electromagnetic spectrum*

YOUNG'S SLITS EXPERIMENT

This consists of two narrow slits placed close together with a bright source of monochromatic (single-wavelength) light behind them. As the light passes through the slits it is diffracted and light from the two sources overlaps. In this region an interference pattern is produced, proving that light is waves rather than particles. More advanced theory and measurements on the pattern produced enables the wavelength of the light to be calculated.

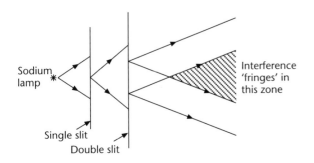

Sodium lamp

Single slit

Double slit

Interference 'fringes' in this zone

The same experiment can be performed with a low-power laser. This allows you to greatly increase the distance from the double slit to the screen so that the interference pattern can be clearly seen on the screen and measured without the need for a magnifying eyepiece.

◆ *Electromagnetic spectrum, Interference*

ZINC-CARBON CELL

This is another name for the dry Leclanché cell that is commonly used in batteries for torches and transistor radios. It is a primary cell producing about 1.5 V and is available in different sizes – a larger cell is able to produce more current if required to do so.

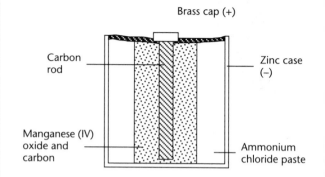

A zinc–carbon dry cell

⊶ Electromotive force

APPENDIX ONE: USE OF UNITS

This appendix will tell you where to find a definition of a unit. This will usually be with a piece of work in which it is commonly used, so you should also find an example of its correct usage. Try to remember that, except for mass, length and time, the name will begin with a 'small' letter but the symbol will be with a capital.

Unit	Symbol	Quantity	Look up
ampere	A	current	**Current**
°celsius	°C	temperature	**Temperature scale**
coulomb	C	charge	**Charge**
decibel	dB	sound level	**Noise**
farad	F	capacitance	**Capacitor**
hertz	Hz	frequency	**Frequency**
joule	J	work/energy	**Work**
kelvin	K	temperature	**Temperature scale**
kilogram	kg	mass	**Mass**
metre	m	length	**Length**
newton	N	force	**Newton's second law**
ohm	Ω	resistance	**Resistance**
pascal	Pa	pressure	**Pressure**
second	s	time	**Time**
volt	V	potential	**Potential difference**
watt	W	power	**Power**

Remember that you can always make the size of a unit larger or smaller by putting one of the prefixes in front of the name of the unit or an extra letter in front of the symbol for the unit.

Prefix	Letter	Unit multiplied by
giga	G	1,000,000,000
mega	M	1,000,000
kilo	k	1,000
unit alone		1
milli	m	1/1,000
micro	μ	1/1,000,000

Examples:

$$1 \text{ megahertz} = 1\,MHz = 1,000,000 \text{ hertz}$$

$$1 \text{ millimetre} = 1\,mm = \frac{1}{1,000} \text{ metre}$$

There are a few *fundamental units* that have exact definitions for the most basic quantities, such as mass, length and time. The other units are formed by carefully combining these together. A unit for speed is formed by combining the units for distance and time – since the definition divides distance by time you also do the same with the units and get m/s. Read the / as 'per'. In the same sort of way, the newton is made up from the units of mass and acceleration, but we use force units so often that it is easier to have a special name for the unit than to use kg m/s each time.

APPENDIX TWO: BASIC EQUATIONS

Here is a quick reference to the basic equations used. The order is alphabetical so that you can find the one that you need more quickly.

$$\text{acceleration} = \frac{\text{change in velocity}}{\text{time taken}} \qquad a = \frac{v - u}{t}$$

Boyle's law:
$$\text{pressure} \times \text{volume} = \text{constant} \qquad P_1 V_1 = P_2 V_2$$

$$\text{charge stored} = \text{capacitance} \times \text{p.d.} \qquad Q = CV$$

$$\text{charge transferred} = \text{current} \times \text{time} \qquad Q = It$$

Charles' law:
$$\frac{\text{volume}}{\text{temperature}} = \text{constant} \qquad \frac{V_1}{T_1} = \frac{V_2}{T_2}$$

$$\text{density} = \frac{\text{mass}}{\text{volume}} \qquad D = \frac{m}{v}$$

$$\text{efficiency} = \frac{\text{energy output}}{\text{energy input}} = \frac{\text{work output}}{\text{work input}}$$

$$\text{efficiency} = \frac{\text{M.A.}}{\text{V.R.}} \times 100\%$$

$$\text{energy} = \text{mass} \times \text{specific heat capacity} \times \text{change in temperature} \qquad W = ms\Delta\theta$$

$$\text{energy changed} = \text{p.d.} \times \text{current} \times \text{time} \qquad W = VIt$$

$$\text{energy changed} = \text{current}^2 \times \text{resistance} \times \text{time} \qquad W = I^2Rt$$

$$\text{expansion} = \text{linear expansivity} \times \text{original length} \times \text{rise in temperature}$$

$$\text{force} = \text{mass} \times \text{acceleration} \qquad F = ma$$

General gas equation:
$$\frac{\text{pressure} \times \text{volume}}{\text{temperature}} = \text{constant} \qquad \frac{P_1 V_1}{T_1} = \frac{P_2 V_2}{T_2}$$

$$\text{gravitational P.E.} = \text{weight} \times \text{height} \times \text{gravitational field strength} \qquad \text{P.E.} = mgh$$

$$\text{heat of fusion} = \text{mass} \times \text{specific latent heat of fusion} \qquad \text{latent heat of fusion} = mL_f$$

$$\text{heat of vaporization} = \text{mass} \times \text{specific latent heat of vaporization} \qquad \text{latent heat of fusion} = mL_v$$

$$\text{kinetic energy} = \tfrac{1}{2} \times \text{mass} \times \text{velocity}^2 \qquad \text{K.E.} = \frac{m \times v^2}{2}$$

$$\text{mechanical advantage} = \frac{\text{load}}{\text{effort}}$$

$$\text{moment} = \text{force} \times \text{perpendicular distance to pivot}$$

$$\text{momentum} = \text{mass} \times \text{velocity}$$

$$\text{potential difference} = \frac{\text{work done}}{\text{charge moved}}$$

$$\text{potential divider equation:} \qquad V_{out} = V_{in} \times \frac{R_1}{(R_1 + R_2)}$$

$$power = \frac{energy\ changed}{time\ taken} = \frac{work\ done}{time\ taken} \qquad P = \frac{W}{t}$$

$$power = p.d. \times current \qquad P = VI$$

$$power = current^2 \times resistance \qquad P = I^2R$$

$$pressure = \frac{normal\ force}{area} \qquad P = \frac{F}{A}$$

$$pressure\ of\ a\ liquid = depth \times density \qquad P = hDg \\ \times gravitational\ field \\ strength$$

pressure law:

$$\frac{pressure}{temperature} = constant \qquad \frac{P_1}{T_1} = \frac{P_2}{T_2}$$

$$resistance = \frac{potential\ difference}{current} \qquad R = \frac{V}{I}$$

$$resistors\ in\ series \qquad R_{total} = R_1 + R_2 + \dots$$

$$resistors\ in\ parallel \qquad \frac{1}{R_{total}} = \frac{1}{R_1} + \frac{1}{R_2} \dots$$

$$speed = \frac{distance\ moved}{time\ taken} \qquad v = \frac{s}{t}$$

$$transformer\ equation: \qquad \frac{V_{out}}{V_{in}} = \frac{N_{secondary}}{N_{primary}}$$

$$velocity = \frac{change\ in\ displacement}{time\ taken} \qquad v = \frac{s}{t}$$

$$velocity\ ratio = \frac{distance\ moved\ by\ effort}{distance\ moved\ by\ load}$$

$$voltage\ gain = \frac{voltage\ output}{voltage\ input} \qquad A = \frac{V_{out}}{V_{in}}$$

$$wave\ velocity = frequency \times wavelength \qquad v = f\lambda$$

$$weight = mass \times gravitational\ field\ strength \qquad weight = mg$$

$$work\ done = force \times distance\ moved \qquad W = Fs \\ in\ direction\ of\ force$$

APPENDIX THREE: CIRCUIT SYMBOLS

This is a list of the circuit symbols that are used by the major examination boards.

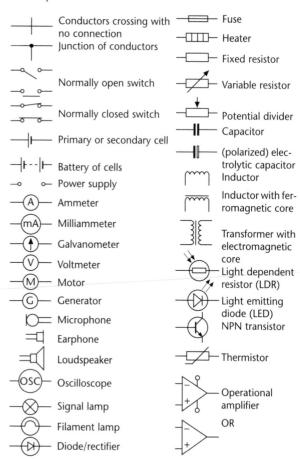

NB: Transistor, divide, LED and LDR may be drawn without the surrounding circle.

Relays

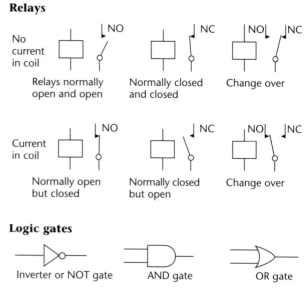

Logic gates

Inverter or NOT gate   AND gate   OR gate

NAND gate   NOR gate

APPENDIX FOUR: DOING AN INVESTIGATION

Your practical investigations will be marked in four sections called Planning (P): 8 marks; Observations (O): 8 marks; Analysis (A): 8 marks; and Evaluation (E): 6 marks. The following will help you to see what you need to do in each section to get the marks. To get the marks for each stage you also need to have done the previous ones (i.e. to get 6 marks you also have to have done the work listed for 2 marks and 4 marks). If you complete all the tasks in a level but do not manage all the things in the next level you may be given a mark between the two.

Planning (P)

For 2 marks:
Describe a simple, safe method of carrying out your investigation. Include a diagram and a list of the apparatus that you use. Write down any safety precautions that you took.

For 4 marks:
Describe what makes your test fair. This usually means that you have to say what things you will need to keep the same.

Make a prediction about what you think will happen – be as exact as you can.

> Remember: Be clear about exactly what it is that you are investigating.

Choose equipment that is suitable, e.g. the correct size of measuring cylinder.

For 6 marks:
Use your *scientific* knowledge to explain why you have chosen the variable that you did and to explain your prediction. Show how you knew which variables had to be controlled.

Decide on a suitable range and number of measurements that you will make, e.g. you might decide to change the length of a pendulum from 0.5 m to 2 m (the range), increasing the length by 0.25 m each time (so the number is fixed).

> Remember: Make sure that you obtain at least five test results so that you can plot a graph, and always make sure that you do repeats to check for accuracy.

For 8 marks:
Extend your scientific reasoning to explain your prediction in more detail and show why you decided on the range and number that you did.

Do preliminary investigations to check all this or back it up with work researched from books, etc. If you use a library, CD-ROM, etc., write a bibliography.

Observations (O)

For 2 marks:
Use your equipment safely and make some observations or measurements.

> Remember: Write down clearly what your results were and what you did to make sure that everything was safe.

For 4 marks:
Make observations or measurements that are suitable for your investigation and are sufficient to arrive at a conclusion. Record the results of your observations or measurements.

For 6 marks:
Your results should be accurate and have been repeated if possible. They should show that you approached them in a systematic manner.

Observations and measurements should be recorded clearly and accurately.

> Remember: Record ALL your results in clear tables, not just the averages that you get from a series of tests. Recording anomalous (rogue) results and their repeats at this stage helps with the evaluation.

For 8 marks:
All your results should have been arrived at with precision and all the equipment used with skill and accuracy.

Your results should have been taken over a reasonably wide range and there should be an appropriate number of them.

Analysis and conclusions (A)

For 2 marks:
Explain simply what you have found out.

For 4 marks:
Present your results in simple diagrams, charts or graphs.

> Remember: This is not the same as putting the results in a table. The least that would usually be expected is a simple bar chart or graph of your results.

Show that you have found a pattern or trend in the results.

For 6 marks:
Process the results to reach a conclusion. This may mean using your results to do calculations to obtain the facts that you need. It will almost always mean that you will have drawn careful graphs with *lines of best fit*.

Draw a conclusion and show how it fits the evidence that you have produced. Support this conclusion by relating it to your *scientific* knowledge and understanding – use the work that you researched for the planning marks.

For 8 marks:
Use *detailed* scientific knowledge and understanding to explain your conclusions. Explain carefully how the results that you obtained supported or undermined the original prediction that you made.

Evaluation (E)

For 2 marks:
Comment about the method that you used or the results that you obtained.

For 4 marks:
Comment on how accurate your results were and show that you spotted any anomalous (rogue) results.

Discuss whether your method was appropriate and suggest improvements to it that would produce more reliable results.

For 6 marks:
Discuss fully the reliability of your results. Explain any anomalous results. Explain whether the evidence that you have obtained is enough to support the conclusions that you arrived at – is further work needed for a firm conclusion? Suggest improvements or further work that would give more evidence for your conclusions or would extend the investigation.

> Remember: You need to put in more detail as you try to get more marks. Some investigations may be too straightforward to allow high marks. Three marks are also available for spelling, punctuation and grammar.

CHECKPOINT ANSWERS

Acceleration

change in velocity $= v - u$
$$= 10 - 25$$
$$= -15 \text{ m/s}$$

$$\text{acceleration} = \frac{\text{change in velocity}}{\text{time taken}}$$

$$= \frac{-15}{3} = -0.5 \text{ m/s}$$

> *Remember: The minus sign tells you that the train is slowing down instead of speeding up.*

Boyle's law

pressure at bottom of lake

$$= \text{pressure of water}$$
$$+ \text{ pressure of air on surface}$$
$$= 2 \text{ atmos.} + 1 \text{ atmos.}$$
$$= 3 \text{ atmospheres}$$

Assuming that the temperature of the water is constant, use Boyle's law:

$$P_1 V_1 = P_2 V_2$$
$$3 \times 2 = 1 \times V_2$$
$$6 = V_2$$
volume at surface $= 6 \text{ cm}^3$

Change of state

The missing words are:

Down the centre – gas, evaporate, melt, freeze.

Clockwise from top left – sublime, condense, liquid, solid.

Charge

charge $=$ current $\times$ time
$$= 0.25 \times 120$$
$$= 30 \text{ C}$$

> *Remember: You MUST use seconds for the time.*

Conduction of heat

Metals: As explained in the text, metals have electrons that can move freely through the solid, taking energy with them.

Good conductors: pans for cooking, metal radiators.
Insulators: polystyrene cups, roof insulation in houses.

There are many correct answers to these last two questions.

Conservation of momentum

Think of the space station as being stationary.

momentum of space station
$= $ mass $\times$ velocity $= 0$

momentum of shuttle
$= $ mass $\times$ velocity
$= 5,000 \times 0.1 = 500 \text{ kgm/s}$

total momentum
$= 500 \text{ kgm/s}$

total mass after collision
$= 35,000 \text{ kg}$

new momentum
$= $ mass $\times$ velocity
$= 35,000 \times$ change in speed

Since momentum is conserved

$35,000 \times$ change in speed $= 500$
change in speed
$$= 500/35,000$$
$$= 0.014 \text{ m/s}$$

Current

current in bulb A $= 0.5 \text{ A}$
(must be the same as in the bulb in series with it)

current in bulb B $= 1.0 \text{ A}$
(total current is 1.5 A – the other 0.5 A passes through the other two bulbs)

Density

$$\text{density} = \frac{\text{mass}}{\text{volume}}$$

$$8,000 = \frac{\text{mass}}{0.25}$$

mass $= 0.25 \times 8,000 = 2,000 \text{ kg}$

This is a complicated sounding way of telling you that 1 kg will weigh 10 N on Earth.

If 1 kg weighs 10 N
2,000 kg weighs 20,000 N

> *Remember: You can get more help on this last bit if you look up **weight***

Distance/ time graphs

(1) $\text{speed} = \dfrac{\text{distance}}{\text{time}}$

$$= \frac{2}{5} = 0.4 \text{ km/min}$$

$$= 24 \text{ km/h}$$

[either answer is OK]

(2) 5 minutes

(3) $\text{average speed} = \dfrac{\text{total distance}}{\text{total time}}$

$$= \frac{4}{20}$$

$$= 0.2 \text{ km/min}$$

$$= 12 \text{ km/h}$$

Efficiency

$$\text{efficiency} = \frac{\text{energy out} \times 100}{\text{energy in}}$$

$$= \frac{5{,}000 \times 100}{8{,}000}$$

$$= 62.5\%$$

Electromotive force The voltages add, so the supply must be 20 × 12 = 240 V

Expansion of solids Machine the axle and wheel so that the axle is a tiny bit too big to fit.
   Then either (1) heat the wheel so that the hole becomes bigger, fit the axle and wait for it to cool, or (2) cool the axle in liquid nitrogen so that it becomes smaller, fit the axle and wait for it to warm up.
   In either case the wheel and axle will grip so tightly that they can be separated only by cutting them apart.

Fuse

power = volts × amps
1,000 = 230 × amps

$$\text{amps} = \frac{1{,}000}{230} = 4.3 \text{ A}$$

This means that a 5 A fuse should be used.

Gravitational potential energy mass of water = volume × density
= 1 × 1,000 = 1,000 kg
weight of water = mass × g
= 1,000 × 10 = 10,000 N
change in gravitational P.E.
= weight × change in height
= 10,000 × 45
= 450,000 J = 450 kJ

Half life

20 min 20 min 20 min 20 min 20 min 20 min
1,600 → 800 → 400 → 200 → 100 → 50 → 25

The total time is six half lives, and the count rate is reduced to 25 counts per second.

> **Remember: (1) The count will go down like this only if the new atom that is formed at each decay is also radioactive. (2) One count per second is also known as 1 Bq (becquerel).**

Heating effect of electric current

power = I²R
= 0.5² × 100
= 0.25 × 100
= 25 W

Hooke's law 200 N produces an extension of 50 mm.

100 N produces an extension of 25 mm.

total length = original length + extension
= 250 + 25 = 275 mm

Kinetic theory

$$\text{average velocity} = \frac{\text{distance moved}}{\text{time taken}}$$

$$= \frac{100}{10} = 10 \text{ m/s}$$

kinetic energy
= ½ × mass × velocity²
= ½ × 75 × 10²
= 3,750 J

Logic gates First draw a truth table.

Inputs		Output LED
Sensor A (dark)	Sensor B (warm)	
0	0	1
0	1	0
1	0	1
1	1	1

This is not one of the tables that you know for the gates and therefore the circuit will need more than one gate. There is also more than one correct answer! Look at the rows when the output is 1. The LED is on when A = 1 or B = 0 (check this carefully).
   This is the same as:

LED = 1 when A = 1 OR NOT B = 1

The NOT (inverter) has been used to make each part of the circuit = 1.
   Now draw the circuit that the statement describes putting input A and input NOT B into an OR gate.

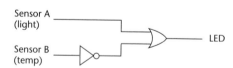

The circuit describes all of the logic part of the circuit, but for those who wish to build and test the answer, the sensor circuits for an LDR and thermistor are included in the following diagram. The other diagram shows an alternative answer to the same problem.

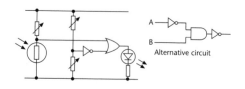

Moment	clockwise moment	*Potential divider*	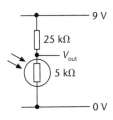

Moment

clockwise moment
= force × perp. distance
= 600 × 2
= 1,200 Nm

anticlockwise moment
= force × perp. distance
= T × 1.5

When board is in equilibrium, anticlockwise moments equal clockwise moments.

T × 1.5 = 1,200
T = 800 N

> Remember: The distances are from the pivot, not from the end of the board.

Newton's second law

1. force = mass × acceleration
    2,000 = 100 × acceleration

    acceleration = $\frac{2{,}000}{100}$ = 20 m/s²

2. change in velocity
    = acceleration × time
    = 20 × 10
    = 200 m/s

3. acceleration
    = $\frac{\text{change in velocity}}{\text{time taken}}$

    = $\frac{0 - 20}{10}$ = $\frac{-20}{10}$

    = −2 m/s²

4. force = mass × acceleration
    = 25,000 × −2
    = −50,000 N

> Remember: The minus sign in the acceleration tells you that the lorry is slowing down. The minus force also tells you that the force is slowing the lorry rather than speeding it up.

Ohm's law

resistance = $\frac{\text{p.d.}}{\text{current}}$

$6 = \frac{12}{1}$

$1 = \frac{12}{6}$ = 2 A

Payment for electrical energy

number of units = kW × hours
= 0.100 × 120 × 6 × 5
= 360 (only 5 days per week!)
cost = units × cost per unit
= 360 × 6 = 2,160p = £21.60

Potential divider

$V_{out} = V_{in} \times \dfrac{R_2}{R_1 + R_2}$

= $9 \times \dfrac{5{,}000}{25{,}000 + 5{,}000}$

= $9 \times \dfrac{5}{30}$ = 1.5 V

Power

work done = force × distance moved
= 2,000 × 20
= 40,000 J

power = $\dfrac{\text{work done}}{\text{time taken}}$

= $\dfrac{40{,}000}{60}$

= 6,667 W = 6.67 kW

Pressure

The drawing pin will have a much greater pressure on the notice board because the force is acting on a much smaller area. The pin can then push its way through the fibres of the board. The area at the point of the pin will be hundreds of times smaller than your thumb, so the pressure from the same force will be hundreds of times greater.

Resistor

Checkpoint 1

47,000 Ω and 5% tolerance.

Checkpoint 2

total resistance = $R_1 + R_2$
= 1,500 + 500
= 2,000 Ω = 2 kΩ

> Remember: 1 kΩ = 1,000 Ω

Resultant forces

1. 500 N to the right.
2. 0 N (forces in equilibrium).
3. 175 N to the right.
4. 400 N to the left.

Sound

The thunder and the light are both created at the same time by the same lightning strike. The light will reach you almost instantly because it travels at such a great speed (300,000,000 m/s). The sound travels more slowly through the air (about 340 m/s) and reaches you later. This will mean that a lightning strike is about 1 km away from you if the thunder arrives about 3 seconds after you see the flash.

Sources of energy

1. coal, oil, natural gas.
2. two from: tidal, geothermal, nuclear.

Specific heat capacity

1. energy = mass × specific heat capacity × change in temperature

$$37,800 = 0.5 \times \text{specific heat capacity} \times 180$$

specific heat capacity

$$= \frac{37,800}{0.5 \times 180}$$

$$= 420 \, \text{J/kgK}$$

2. energy = mass × specific heat capacity × change in temperature
$$= 30 \times 500 \times 480$$
$$= 7,200,000 \, \text{J} = 7.2 \, \text{MJ}$$

Speed

$$\text{speed} = \frac{\text{distance}}{\text{time taken}}$$

$$= \frac{225}{15} = 15 \, \text{m/s}$$

Static electricity

The tiny charged drops will repel each other and spread out. They will be attracted to the earthed leaves of the plant and will coat both sides of the leaves. Larger drops with no charge would only fall onto the top of the leaves and often run off. The spray also has a wider area of coverage because the drops spread out, and it is less wasteful.

Temperature scales

273 K, 373 K, 300 K, 293 K, 263 K, 546 K.

Transformer

$$\frac{V_{\text{out}}}{V_{\text{in}}} = \frac{\text{secondary turns}}{\text{primary turns}}$$

$$\frac{2}{230} = \frac{\text{secondary turns}}{460}$$

secondary turns = 4

U-values

area of window = 1 × 2 = 2 m²
temperature difference
= 20 − 5 = 15 °C
power loss through window
= U-value × area × temperature difference
= 3.0 × 2 × 15 = 90 W

area of wall = (2.5 × 4) − (1 × 2)
= 10 − 2 = 8 m²

temperature difference
= 20 − 5 = 15°C
power loss through wall
= U-value × area × temperature difference
= 1.7 × 8 × 15 = 204 W

total power loss
= 204 + 90 = 294 W

Velocity/ time graphs

Checkpoint 1

2 m/s², 0 m/s², 1.5 m/s², −7 m/s².

> Remember: The minus sign in the last answer means that the object is slowing down rather than speeding up.

Checkpoint 2

75 m, 100 m, 275 m, 87.5 m.

If you have not got these answers check that you used the numbers from the scales and that you did not just count the squares.

total distance moved in 25 s
= total area under the graph
= 75 + 100 + 275 + 87.5
= 537.5 m

Wave

1. Transverse (light/water ripples) and longitudinal (sound).
2. Vibration in a transverse wave is across the direction of travel. In a longitudinal wave the vibration is for and against the direction of the wave.
3. Energy.

Wave equation

velocity = frequency × wavelength
= 250 × 1.36 = 340 m/s

The velocity of the sound wave will stay the same so:

velocity = frequency × wavelength
340 = 200 × wavelength

$$\text{wavelength} = \frac{340}{200} = 1.7 \, \text{m}$$

Work

Remember that you need to deal with force (weight) and not mass, so:

weight of boy = mass × g
= 60 × 10 = 600 N

work done = force × distance moved
= 600 × 10
= 6,000 J = 6 kJ

> Remember: Doing this will transfer the energy that the boy has taken in. Chemical energy in his food has become gravitational potential energy.